ADVANCES IN ENZYMOLOGY

AND RELATED AREAS OF
MOLECULAR BIOLOGY

Volume 76

ADVANCES IN ENZYMOLOGY

AND RELATED AREAS OF MOLECULAR BIOLOGY

Edited by ERIC J. TOONE

DUKE UNIVERSITY, DURHAM, NORTH CAROLINA

WILEY

A JOHN WILEY & SONS, INC., PUBLICATION

Published by John Wiley & Sons, Inc., Hoboken, New Jersey
Published simultaneously in Canada

Library of Congress Cataloging-in-Publication Data:

ISBN 978-0471-23584-2

Printed in the United States of America

10 9 8 7 6 5 4 3 2 1

CONTENTS

CONTRIBUTORS

JOSEPH A. CHEMIER, Department of Chemical and Biological Engineering, University of Buffalo, 303 Fumas Hall, Buffalo, NY 14260

ZACHARY L. FOWLER, Department of Chemical and Biological Engineering, University of Buffalo, 303 Fumas Hall, Buffalo, NY 14260

ERIK HOLTZAPPLE, Department of Biochemistry, Molecular Biology and Biophysics, University of Minnesota, 1479 Gortner Avenue, St. Paul, MN 55108

ETHAN T. JOHNSON, Department of Biochemistry, Molecular Biology and Biophysics, University of Minnesota, 1479 Gortner Avenue, St. Paul, MN 55108

*MATTHEOS A. G. KOFFAS, Department of Chemical and Biological Engineering, University of Buffalo, 303 Fumas Hall, Buffalo, NY 14260 [email: mkoffas@eng.buffalo.edu]

EFFENDI LEONARD, Department of Chemical Engineering, Massachusetts Institute of Technology, 77 Massachusetts Avenue, Amherst, MA 02139

*GEORGE D. MARKHAM, The Fox Chase Cancer Center, Institute for Cancer Research, 333 Cottman Avenue, Philadelphia, PA 19111 [email: gd_markham@fccc.edu]

MAXIM PIMKIN, The Fox Chase Cancer Center, Institute for Cancer Research, 333 Cottman Avenue, Philadelphia, PA 19111

*CLAUDIA SCHMIDT-DANNERT, Department of Biochemistry, Molecular Biology and Biophysics, University of Minnesota, 1479 Gortner Avenue, St. Paul, MN 55108 [email: schmi232@umn.edu]

*JON S. THORSON, Laboratory for Biosynthetic Chemistry, Pharmaceutical Sciences Division, School of Pharmacy, National Cooperative Drug Discovery Program, University of Wisconsin—Madison, 777 Highland Avenue, Madison, WI 53705 [email: jsthorson@pharmacy.wisc.edu]

GAVIN J. WILLIAMS, Laboratory for Biosynthetic Chemistry, Pharmaceutical Sciences Division, School of Pharmacy, National Cooperative Drug Discovery Program, University of Wisconsin—Madison, 777 Highland Avenue, Madison, WI 53705

* Corresponding author.

PREFACE

Natural products continue to play an important role in the treatment of human disease. Among the most daunting limitations to the use of such species is the need for synthetic approaches to the significant quantities of material needed for biochemical study, preclinical and clinical evaluation and, ultimately, supply. Although synthetic chemistry has made tremendous advances in enantiopure synthesis, material limitations still loom large in the development of natural product therapeutics.

Enzymes are by now well established as chiral catalysts for the construction of small chiral synthons, and the use of various esterases, lipases, oxidoreductases, and other enzymes is well known both for the resolution of racemates and for the desymmetrization of optically inactive meso compounds. But the construction of complex natural products often requires dozens of transformations beginning from such simple starting materials. Complex materials are constructed *in vivo* through multiple sequential enzymatic transformations, producing products of startling complexity with apparent ease. The development of novel *in vivo* biosynthetic pathways might ease the material limitations for drug development: the construction and use of such pathways forms the focus of this volume of *Advances in Enzymology.*

A surprising number of biosynthetic pathways are modular, creating a diversity of products by subtle rearrangement of genes. This modular approach to natural product biosynthesis provides a potentially powerful approach to the construction of large numbers of related complex natural products and to the preparation on scale of single compounds. Two chapters in this volume, by Claudia Schmidt-Dannert and coworkers and by Mattheos Koffas and coworkers, review various aspects of such approaches. Glycosylation is increasingly recognized as a powerful modulator of the biological activities of complex natural products, both with regards to biological lifetimes and affinity and avidity at the target receptor. Chemical glycosylation is notoriously refractive, and *in vivo* approaches to glycosylation are especially attractive: the use of native and engineered glycosyl transferases for the preparation of glycosylated natural products is described here by Williams and Thorson. Finally, Pimkin and Markham describe the structure, activity and inhibition of the nicotinamide-dependent inosine-5′-monophosphate dehydrogenase.

ERIC J. TOONE

INOSINE 5'-MONOPHOSPHATE DEHYDROGENASE

By MAXIM PIMKIN and GEORGE D. MARKHAM, *The Fox Chase Cancer Center, Institute for Cancer Research, Philadelphia, Pennsylvania, 19111*

CONTENTS

Purine nucleotide biosynthesis has been an enticing area for biochemists and medicinal chemists since the pathways began to be elucidated half a

Advances in Enzymology and Related Areas of Molecular Biology, Volume 76
Edited by Eric J. Toone   Copyright © 2009 by John Wiley & Sons, Inc.

century ago (1). The central roles of purine nucleotides in energy metabolism, signal transduction, genetic information storage and translation have resulted in a vast literature on these systems. The variety of chemical transformations involved in the transmogrification of ammonia, amino acids, and CO_2 into the heterocyclic purines continues to lure mechanistic enzymologists to this area. Inosine $5'$-monophosphate dehydrogenase [IMP-NAD oxidoreductase, Enzyme Classification (E. C.) 1.2.1.14; IMPDH] catalyzes a metabolic branch point reaction in purine synthesis and has been an area of intellectual convergence in biological and medicinal chemistry. The intent of this chapter is to discuss the state of knowledge of the mechanism of IMPDH in the context of the intimate relation of mechanism to protein structure. For a recent minireview encompassing much of the progress made in the last five years see ref. 2. The immense contribution of medicinal chemistry to design of inhibitors and their clinical utility is beyond the scope of this chapter (cf. refs. 3 and 4), but the mechanistic insights provided by selected inhibitors are incorporated. We hope that some of the outstanding questions regarding IMPDH structure and function are highlighted herein.

Figure 1 illustrates that IMP is located at the metabolic branch point wherein this purine becomes committed to either guanosine monophosphate (GMP) or adenosine monophosphate (AMP) synthesis. IMP may originate from de novo purine biosynthesis, be salvaged from AMP catabolism, or be formed from hypoxanthine and 5-phosphoribosyl-1-pyrophosphate (PRPP).

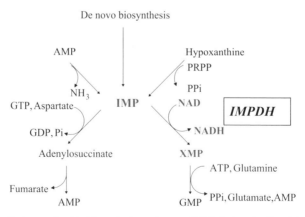

Figure 1.   Purine nucleotide biosynthetic pathways: PRPP, 5-phosphoribosyl-1 pyrophosphate; XMP, xanthosine 5-monophosphate.

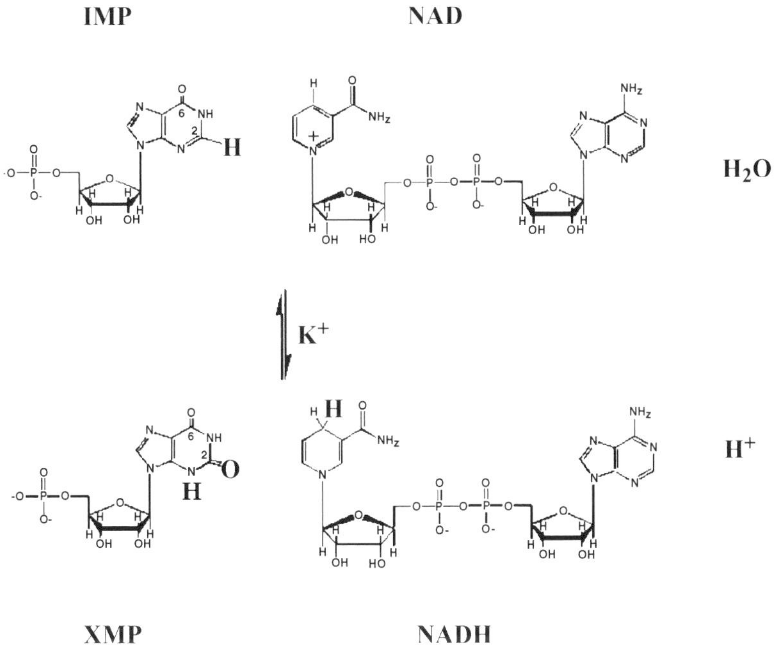

IMP **NAD**

Figure 2.   IMPDH-catalyzed reaction.

A schematic of the IMPDH reaction is shown below, and the chemical structural changes are illustrated in Figure 2.

$$IMP + NAD + H_2O \xrightarrow{K^+} XMP + NADH + H^+$$

The IMPDH-catalyzed synthesis of XMP is the rate-limiting step in the biosynthesis of guanine nucleotides in many organisms (1, 5). Despite this pivotal metabolic location, allosteric regulation of IMPDH has not been substantiated. Modest inhibition by the end products of purine biosynthesis, AMP and GMP, has often been observed but the in vivo contribution of these interactions is typically unclear (6). In mammals, which typically have two IMPDH isozymes, regulation at the level of gene expression has been characterized (7, 8). Mice that are homozygotic knockouts for the gene that encodes the type 2 isozyme die in early embryonic stages, despite the presumptive expression of a functional type 1 isozyme and purine salvage enzymes (7). Thus, there appears to be an important role of the type 2 isozyme in mammalian development. IMPDH expression in mammals is controlled by the central metabolic regulator p53, which is the most commonly mutated protein in human cancers (9, 10). The enhanced expression of the type 2 isozyme in leukemias and other proliferating human cells has long been recognized (11). Recent work suggests that the expression of both isozymes increases under these conditions (8, 12). Diverse pharmacological applications for IMPDH inhibitors stem from the reliance of proliferating lymphocytes, and some parasites, on IMPDH activity.

I. Overview of IMPDH

Prominent features of IMPDH include the following:

1. The IMPDH monomer is composed of a core $\sim$400-residue $(\alpha\beta)_8$ barrel where the IMPDH reaction occurs. The protein from most organisms has an evolutionarily conserved $\sim$120-residue domain of unknown function that is appended to the α_2–β_3 junction of the barrel. The subunits form a homotetramer. Loops between α/β segments contain catalytically important residues; these loops are often crystallographically disordered, which apparently reflects significant conformational alterations associated with catalysis.

2. The kinetic mechanism has random substrate addition with a preferred path of IMP being the first substrate to bind. A conformational change appears to follow IMP binding. Product release is ordered, with XMP the last to dissociate. This is in contrast to most dehydrogenases in which the NAD(H) binds first and its counterpart exits last. XMP release rather than a chemical reaction is rate limiting.

3. The chemical mechanism proceeds through a covalent adduct between the 2-posititon of the IMP purine ring and the sulfur of an active-site cysteine. Hydride transfer from the covalent enzyme–IMP species to NAD yields a thioimidate (denoted E-XMP*) that is then hydrolyzed. This mechanism has similarities to that of glyceraldehyde 3-phosphate dehydrogenase (13). Following hydride transfer and NADH release, a mobile loop structure, called the "flap," moves into the vacant NADH site and its residues activate water for E-XMP* hydrolysis.

4. A monovalent cation such as K^+ is required for maximal dehydrogenase activity and appears to be related to conformational stabilization at the NAD site. However, its detailed roles remain elusive.

IMPDH has been purified and characterized from a plethora of organisms beginning with the studies of Magasanik and co-workers in *Aerobacter aerogenes* (14). This chapter focuses on examples of IMPDH that have been characterized in substantial enzymatic and structural detail, with particular emphasis on those for which three-dimensional crystal structures are available. Functional properties of these representatives are listed in Table 1. It is notable that the K_m values for IMP are comparable among the IMPDH from various organisms, typically near 10^{-5} M. In contrast, the K_m for NAD varies more than 50-fold, with larger values for the bacterial enzymes. All known IMPDHs require a monovalent cation such as K^+ for maximal activity, and the still uncertain role(s) of this cation is discussed below. Table 1 also includes inhibition constants for selected compounds that have been of utility in elucidating the structure or mechanism. The therapeutic precursors of these inhibitors are illustrated in Figure 3. Mizoribine and ribavirin are metabolized to the monophosphate IMP analogs. Tiazofurin is converted in vivo to the NAD analog tiazofurin adenine dinucleotide (TAD); some studies have used the nuclease-resistant β-methylene-bisphosphonate TAD analog (β-TAD). Mycophenolic acid (MPA) is itself the active form.

A. PROTEIN STRUCTURES

IMPDH from eucarya, bacteria, and archaea is typically composed of ~500 amino acids. Notably, the IMPDH from *Borrelia burgdorferi* (the causative agent of Lyme disease) lacks an appended domain and is ~100 residues smaller than most other IMPDH (vide infra). IMPDH is fairly highly conserved in sequence throughout nature; for example, the two human IMPDH isozymes have 84% sequence identity through their entire 514 amino acids, and the hamster and human type 2 enzymes vary in only 6 residues; eucaryotic and microbial IMPDHs typically share ~30% identity. A phylogenetic analysis indicated a deep branching of bacterial and eucaryotic IMPDH (29).

IMPDH proteins generally form tetramers, but the protein isolated from most species has been noted to aggregate (a property that has frustrated spectroscopic studies). Crystal packing shows that the tetramers can stack on top of one another and this may be related to the gross aggregation that limits

TABLE 1
Kinetic Properties of IMPDH from Selected Organisms

| Organism | k_{cat} (s^{-1}) | K_m | | K_a, K$^+$ (mM) | K_i | | | | Reference |
		IMP (μM)	NAD (mM)		MMP (μM)	RMP (nM)	MPA (μM)	TAD (μM)	
Human-1	1.8	14	42	0.65	8.2	0.65	11	0.095(β)	15
Human-2	1.4	9.2	32	0.39	3.9	0.39	6	0.14(β)	15, 16
Tritrichomonas foetus	1.9	1.7	150	12	NR	0.065	9,000	2.3(β)	17–20
Escherichia coli	13	60	2,000	2.8	0.5	NR	NR	8.5(β)	21
Borrelia burgdorferi	2.6	29	1,100	25	NR	NR	8,000	1.6	22
Streptococcus pyogenes	24	62	1,180	NR	0.5	6	>10,000	NR	23
Cryptosporidium parvum	3.3	29	150	NR	11	NR	NR	1.5(β)	24

Abbreviations: MMP, mizoribine monophosphate; RMP, ribavirin monophosphate; MPA, mycophenolic acid; TAD, tiazofurin adenine dinucleotoide. See Figure 3 for structures. NR: not reported; β: β-CH$_2$-TAD was used.

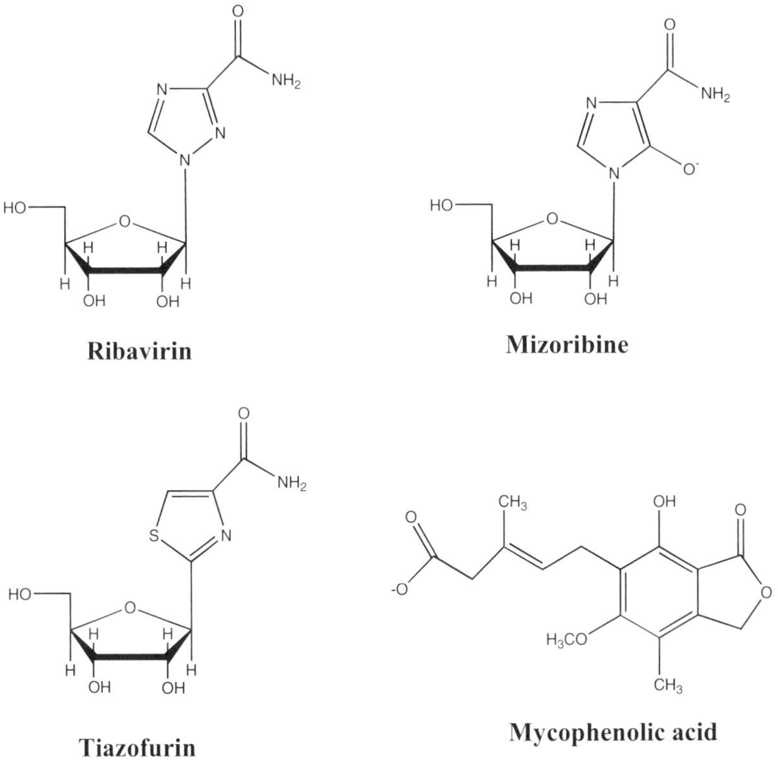

Ribavirin

Mizoribine

Tiazofurin

Mycophenolic acid

Figure 3. Structures of chemotherapeutic agents that have had influence on structural and mechanistic studies of IMPDH. Ribavirin and mizoribine are active as the nucleoside monophosphates that bind at the IMP site (25, 26). Tiazofurin is activated to the NAD analog tiazofurin adenine dinucleotide (27). Mycophenolic acid binds at the nicotinamide portion of the NAD site (28).

solubility. The role of the K^+ is not to maintain an active tetrameric state, at least for the human type 2, *B. burgdorferi*, *Escherichia coli*, or *Tritrichomonas foetus* enzymes; for example, the human enzyme remains tetrameric in the presence of the noninteracting $(CH_3)_4N^+$ ion (16, 30).

Crystal structures have been reported for IMPDH from hamster (type 2), human (types 1 and 2), bovine parasite *T. foetus*, and bacteria *B. burgdorferi* and *Streptococcus pyogenes* (19, 20, 23, 28, 31–35). The variety of ligands in the structures enhances the functional information that can be gleaned from them. Selected information from the crystal structures is listed in Table 2.

TABLE 2
Crystal Structures of IMPDH

Organism	Resolution (Å)	Ligands	Reactive Cysteine Number	Protein Data Bank Code	Reference
Hamster	2.6	XMP*, MPA	331	1jr1	28
Human-1	2.6	6-Cl-IMP	331	1jcn	32
Human-2	2.9	6-Cl-IMP + SAD	331	1b3o	31
	2.9	6-Cl-IMP + NAD	331	1nfb[a]	
	2.65	Ribavirin-MP + C2-MAD[b]	331	1nf7[a]	
B. burgdorferi	2.4	sulfate	229	1eep[a]	34
S. pyogenes	1.9	IMP	310	1zfj	23
T. foetus	2.3	none	319	1ak5	33
	2.6	XMP	319	—[c]	33
	1.9	Ribavirin-MP	319[d]	1me8	19
	2.5	Ribavirin-MP + MPA	319[d]	1me7	19
	2.2	IMP	319	1me9	36
	1.95	IMP + MPA	319	1meh	36
	2.2	XMP + MPA	319	1mei	36
	2.15	NAD+ + XMP	319	1mew	36
T. foetus Δ(101–226)	2.2	IMP + β-CH₂-TAD	319	1lrt	20
	2.0	MMP	319	1pvn	35

[a]Unpublished.
[b]C2-MAD, C2-mycophenolic adenine nucleotide.
[c]Not deposited.
[d]In these ribavirin-MP structures, the active-site cysteine had oxidized to the sulfinic acid.

The structure of the IMP complex of *S. pyogenes* IMPDH protein has the highest available resolution of 1.9 Å (23) and is illustrated in Figure 4A. Figure 4B shows the structure of the human type 2 isozyme covalently modified by the affinity label 6-chloropurine riboside 5′-monophosphate (6-Cl-IMP) and complexed with the NAD analog selenazole adenine dinucleotide (SAD) (31). The different overall shape in the figure reflects the larger regions of disorder in the human enzyme structure.

The sequences of these IMPDH, and the well-studied *E. coli* enzyme, are illustrated in Figure 5. The substantial segments of the proteins that are disordered in the crystal structures are denoted by the crossed-out residues. The relationship between the evident protein flexibility and catalytic activity is an ongoing area of investigation.

The protein consists of a core ∼400-residue $(\beta/\alpha)_8$ barrel that is structurally highly conserved and is responsible for the IMPDH activity. A subdomain or flanking domain is inserted between the $\alpha_2-\beta_3$ elements of the barrel in all but the *B. burgdorferi* enzyme; hence the cognate residue numbers in the *B. burgdorferi* protein are smaller by ∼90 after residue 96 (see Figure 5). The core domain has dimensions of approximately $40 \times 40 \times 50$ Å. The $(\beta/\alpha)_8$ barrel fold is common to many proteins, for example, triose phosphate isomerase (TIM), pyruvate kinase, and GMP reductase (37). This is an apparently rigid scaffold upon which functional elements are grafted between secondary structural elements. Flexible segments, which are poorly defined in many of the IMPDH structures, connect the barrel components and have been called "loops" and flaps. Particularly important are the loop containing the catalytic cysteine, which connects the $\beta_6-\alpha_6$ segments of the barrel (residues 326–342 in the human enzymes), and the larger flap between the $\beta_8-\alpha_8$ barrel segment (residues 400–450 in the human enzymes). Figure 6A shows an overlay of a single subunit from the structures of the hamster, human type 2, *S. pyogenes*, *B. burgdorferi*, and *T. foetus* proteins. The comparison illustrates the spatial location of these sequence elements (which are boxed in Figure 5) as well as the structural variability seen in crystals.

The active-site cysteine is contained in loop 6, which is a mobile section of the protein periphery connecting β_6 and α_6 secondary structural elements. The loop is relatively conserved in length, but the sequence conservation is less strict. In contrast, the analogous loop in TIM, which is perhaps the most extensively studied active-site loop, is highly conserved in length and sequence (38). In the *T. foetus* apo enzyme and E·XMP

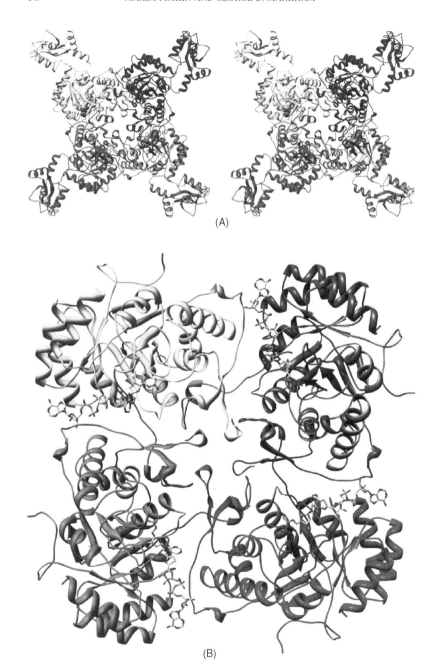

(A)

(B)

complex structures, the disordered region encompassed the glycine-rich loop (residues 313–330), including the active-site cysteine 319 (33). In the newer *T. foetus* structures, loop residues 313–330 are ordered (19, 20). It appears that *T. foetus* active-site loop conformation is stabilized only when the active site is occupied, since it cannot be visualized in the apo enzyme (36). *T. foetus* Gly315 is substituted with a Pro in the *B. burgdorferi* IMPDH, likely providing for an increased loop stability and better visibility.

The "flap" comprises residues numbered ∼400–450 in most sequences that are inserted between the β_8–α_8 barrel segments. The sequence of this flap is the least conserved region of the active site. Even in the 1.9-Å resolution structure of the *S. pyogenes* enzyme, the flap residues 396–419 showed low density, indicating disorder. Because the flap is presumed to close over the active site during catalysis, it was surprising that in the complex of the *T. foetus* enzyme with IMP and β-TAD the flap remains largely disordered despite occupation of all of the binding sites (20). In contrast, the entire flap was well visualized in the *T. foetus* E·MMP complex by Gan et al. (35). The distal portion of the flap interacts with the NAD^+ binding site, creating a previously unobserved closed active-site conformation (see Figure 6A). Gan et al. proposed that the movement of the flap in and out of the NAD site is the necessary conformational change converting the enzyme from a dehydrogenase to a hydrolase.

A remarkable but unexplained phenomenon is the paucity of tryptophan in IMPDH proteins from archaea, bacteria, and eucarya—most available IMPDH sequences have no tryptophan or a single tryptophan residue. It would seem unlikely that this deficit would be coincidental across evolution, but no rationale is obvious. The fluorescence from the multiple tryptophans naturally present in the *T. foetus* IMPDH has been fruitfully used by Hedstrom and co-workers in their mechanistic studies (vide infra) (18).

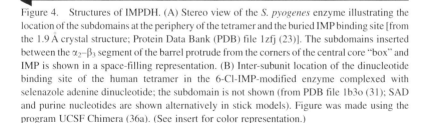

Figure 4.   Structures of IMPDH. (A) Stereo view of the *S. pyogenes* enzyme illustrating the location of the subdomains at the periphery of the tetramer and the buried IMP binding site [from the 1.9 Å crystal structure; Protein Data Bank (PDB) file 1zfj (23)]. The subdomains inserted between the α_2–β_3 segment of the barrel protrude from the corners of the central core "box" and IMP is shown in a space-filling representation. (B) Inter-subunit location of the dinucleotide binding site of the human tetramer in the 6-Cl-IMP-modified enzyme complexed with selenazole adenine dinucleotide; the subdomain is not shown (from PDB file 1b3o (31); SAD and purine nucleotides are shown alternatively in stick models). Figure was made using the program UCSF Chimera (36a). (See insert for color representation.)

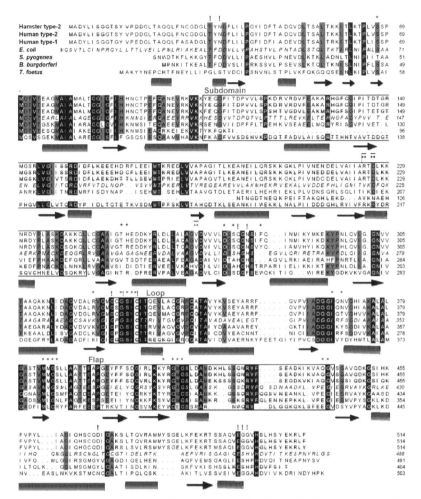

Figure 5. Sequence comparisons for most extensively discussed IMPDH. These are the hamster type 2 enzyme, the two human isozymes, and those from *S. pyogenes*, *E. coli*, *B. burgdorferi*, and *T. foetus*. Residues that are identical in all sequences are shown on a black background, while those identical in five of the six sequences are on light grey. For the five proteins for which crystal structures are known, the residues that are not reported in the PDB files are crossed out; a crystal structure of the *E. coli* enzyme has not been reported. The secondary-structure representation is taken from the *S. pyogenes* structure and appears to be representative of the other IMPDH. The residues that interact with IMP and NAD are marked with single asterisks and those that bind K$^+$ by exclamation points. The three residues initially found mutated in patient families with RP10 retinitis pigmentosa are designated with double asterisks. The active-site loop, flap, and subdomain are boxed.

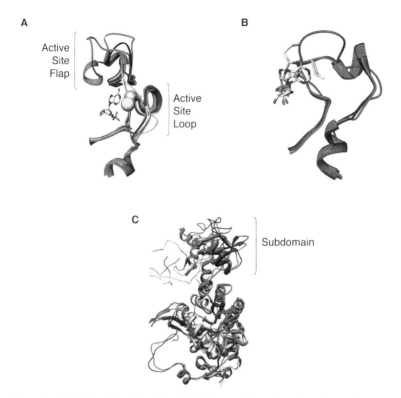

Figure 6. IMPDH conformational flexibility. Superposition of α-carbon traces of monomer core domains in different structures. (A) Structural flexibility of the active-site loop and active-site flap. Superimposition of the entire core domains of 1jr1, 1b3o, 1eep, 1zfj, 1ak5, and 1pvn. The substrate IMP of *S. pyogenes* structure (1zfj) is shown in CPK-colored stick representation. The 1zfj active-site Cys is shown as a CPK-colored space-fill model. (B) Change in the active-site loop 6 conformation upon 6-Cl-IMP binding. Superimposition of the entire core domains of 1nfb, 1zfj, and 1jr1. The IMP substrate of 1zfj and the 6-Cl-IMP adduct are shown in stick representation. (C) Position of the CBS subdomain in different IMPDH structures relative to the enzyme core domain. The substrate IMP of 1zfj structure is illustrated in red space-fill representation. The human type 2 structure is in gold (1b3o), the hamster structure in dark blue (1jr1), the *S. pyogenes* (1zfj) structure in green, the *B. burgdorferi* structure (1eep) in grey, the *T. foetus* apo enzyme structure in magenta (1ak5), the human type 2 structure in complex with 6-Cl-IMP and NAD in red, and the *T. foetus* E·MMP complex in cyan (1pvn). UCSF Chimera (36a) was used for the coordinate superposition and structure visualization. (See insert for color representation.)

II. Medicinal Applications of IMPDH Inhibitors

Purine nucleotide biosynthetic enzymes have long been targets for the development of chemotherapeutic agents (3). As a key enzyme in guanine biosynthesis, the history of IMPDH is replete with searches for analogs of potential clinical utility (5, 11, 25). For a review of cofactor mimics as IMPDH inhibitors see refs. 4 and 39. The recognition that IMPDH type 2 expression is enhanced in growing mammalian cells, such as tumor cells and proliferating lymphocytes, directed substantial attention to this enzyme as a target for inhibitory cancer chemotherapeutic and immunosuppressive agents (11). However, selective inhibition of IMPDH type 1 was recently shown to suffice for endothelial cell growth arrest resulting in inhibition of tumor angiogenesis (40). This observation contrasts with the predominant view that development of selective IMPDH type 2 inhibitors will improve their clinical utility. It is noteworthy that clinically efficacious drugs have been found that interact with either the IMP site or the NAD site (25, 41). Figure 3 illustrates some of the compounds that have had particular impact on structural and mechanistic studies.

The structurally related compounds mizoribine, ribavirin, and tiazofurin form different active metabolites: mizoribine and ribavirin are phosphorylated in vivo to $5'$-monophosphates that act as IMP analogs. Tiazofurin is intracellularly converted to TAD, a NAD analog inhibitor (25). MPA is an uncompetitive inhibitor that binds at the nicotinamide portion of the NAD site. MPA traps the covalent enzyme–XMP* intermediate during turnover; thus this key compound would not have been found in a typical screen for compounds that simply bind with high affinity.

IMPDH inhibitors in clinical use include a pro-drug of mycophenolic acid (CellCept™) and mizoribine (bredinin) for immunosuppression in transplantation (26). Ribavirin is used in combination therapy for hepatitis C infections (42, 43). In cancer chemotherapy, attention has been devoted to the thiazole C-nucleoside tiazofurin (44–47). Tiazofurin not only is toxic to rapidly growing cells but also induces differentiation (48). In contrast, the 6-mercaptopurine used in chemotherapy is converted by IMPDH to 6-thio-XMP in a step toward formation of the toxic 6-thio-guanosine nucleotides that are incorporated into nucleic acids (3). IMPDH has also been a target for development of anti protozoan agents and numerous studies have investigated the structure and mechanism of IMPDH from the bovine pathogen *T. foetus* (17–20, 49–53).

A.   IMP ANALOGS

IMP dehydrogenase has been a classic test system for nucleotide analogs as alternate substrates and inhibitors (25, 54–58). Interest in IMP analogs stems both from pharmacological importance of IMPDH and the flavour-enhancing properties of IMP itself (59). IMPDH was investigated in some of the early studies of affinity labeling by purine nucleotides using 6-chloro and 6-mercapto analogs of IMP (60, 61). These studies showed that inactivation by the 6-chloro analog of IMP, 6-Cl-purine riboside 5-monophosphate (often called 6-Cl-IMP in the context of IMPDH), resulted from reversible complex formation followed by modification of a sulfhydryl group (17, 62, 63). The active-site location and functional importance of this cysteine residue, later identified as cysteine 331 in human type 2 IMPDH, have been amply documented (64). A variety of IMP analogs have been tested as substrates and inhibitors of the human type 2 enzyme (30).

Detailed kinetic studies of mizoribine monophosphate (MMP) inhibition indicate that it acts as a transition state analog inhibitor (65). This behavior was deduced from the proportionality of the inhibition constant of MMP to the value of $k_{cat}/[K_m(IMP)*K_m(NAD)]$ for a series of mutants of the E. coli enzyme and further supported by a recent crystal structure of T. foetus E·MMP complex (35). MMP, the active-site Cys319, and a water molecule form a tetrahedral geometry, characteristic of a transition state for the hydrolysis of a covalent intermediate (Figure 7). The variation in affinity of MMP for different IMPDH illustrates the potential for development of species-selective inhibitors (Table 1). A purine ring expanded ("fat base") nucleoside imidazo[4, 5-e]diazapine, in which an additional methylene group creates a seven-member ring in place of the six-member system of XMP, is a potent slow-binding inhibitor of the E. coli and human enzymes; the kinetics may reflect formation of a covalent adduct which is labile upon protein denaturation (66).

B.   NAD ANALOGS

The efforts to develop clinically useful NAD analogs with selectivity for IMPDH have been extensively reported (4, 67). Tiazofurin and its derivatives such as the selenium analog selenazofurin have been lead compounds (27, 45, 46, 68, 69). Additionally, the natural product MPA binds at the nicotinamide part of the NAD site (52, 70). MPA has recently been demonstrated to induce

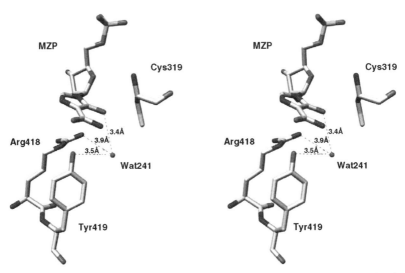

Figure 7.   Stereoview of the transition state analogy of MMP in *T. foetus* structure (1pvn) and mechanism of water activation. The distances are indicated with dotted lines. (See insert for color representation.)

reversible aggregation of IMPDH in vitro and in vivo (71). These enzyme conglomerates are not associated with cellular organelles and the aggregation is partially reversed by GTP. The functional significance of these observations is unclear (71). The in vivo metabolic inactivation of the inhibitory forms of both tiazofurin and MPA has compromised clinical efficacy (72, 73). Recent studies have prepared related compounds that are metabolically stable. These include β-CH$_2$-TAD, which has a methylene bisphosphonate linkage of the two ribotide components, the benzimidazole analog of NAD (BAD), and a conjugate of MPA and adenosine monophosphate, denoted MAD (47).

C.   NATURAL PRODUCT INHIBITORS

These include the aforementioned mycophenolic acid reported in 1896 (74) and recently modified into a clinically useful compound, and the algal product isorawsonol (75). The "fat base" nucleoside developed by Hedstrom and co-workers (66) is related to the natural product azepinomycin (76), and the authors remarked that the ribotide of azepinomycin might be a potent inhibitor of IMPDH.

D. NOVEL SYNTHETIC INHIBITORS

Recently a spate of novel inhibitory compounds of pharmaceutical interest have been reported by both academic and commercial groups; a discussion of these compounds, which are structurally not closely related to the natural ligands, can be found in the references (77–81). A first structure-based IMPDH inhibitor, VX-497, has been developed by Vertex Pharmaceuticals (77). High-throughput screening has been employed at Bristol-Meyers Squibb for discovery of a novel acridone-based IMPDH inhibitor (82).

III. Kinetic Mechanism and Substrate Interactions

The kinetic mechanism of IMPDH has been studied for decades (18, 51, 62, 83–89). A schematic of the IMPDH-catalyzed mechanism is shown in Figure 8A. The binding of the water substrate is not illustrated nor is

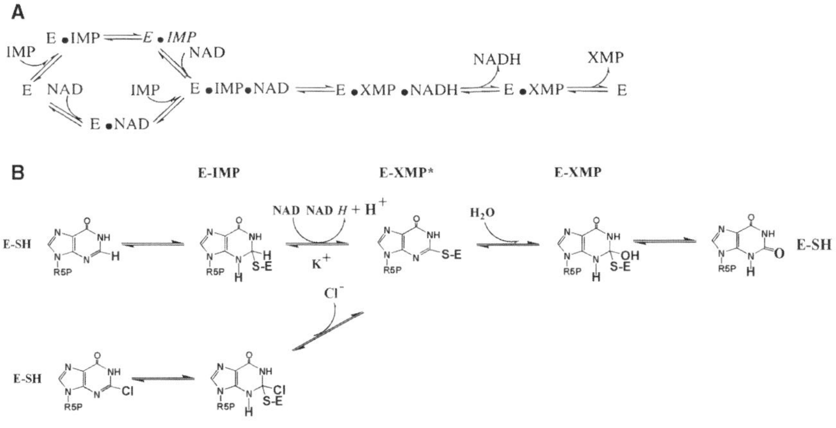

Figure 8. (A). IMPDH kinetic mechanism deduced from steady-state kinetics, isotope exchange, kinetic isotope effects, and pre-steady-state studies. The isomerized E·IMP complex is shown in italics. Noncovalent complexes are connected by "·". The point at which the H_2O substrate binds is not known, nor is whether the K^+ dissociates during each turnover; these participants are not included. (B) Chemical transformations of the enzyme-bound species during the IMPDH reaction. The covalent enzyme adduct is shown as E-XMP*. This scheme includes the putative tetrahedral intermediates E-IMP and E-XMP that have not been experimentally observed. The initial noncovalent (Michaelis) complexes are not shown. The potential confluence of the 2-Cl-IMP hydrolytic reaction pathway is illustrated. The ribose 5'-monophosphate of the substrate is abbreviated as R5P.

the K^+ ion(s) activator. The E·XMP·NAD complex involved in substrate inhibition by NAD is not included. It is reassuring that studies of IMPDH from various sources (both organisms and laboratories) have generally provided a consistent view of the mechanism; thus the presentation below attempts to integrate the salient points of various studies while pointing out their individuality. Key findings are the random binding order with preference for IMP binding first, ordered product dissociation with XMP dissociating last, and the effective irreversibility of the overall reaction.

The results of Heyde and Morrison with the A. aerogenes enzyme revealed the random binding of IMP and K^+, with NAD associating to enzyme·K^+ complexes (83, 84). They proposed that binding of K^+ was linked to formation of the NAD binding site, a concept that has become a consistent theme over the decades. The authors presented a cogent discussion of influence of the substrate concentrations employed on the apparent kinetic mechanism because of differences in binding rates. At a saturating K^+ concentration there is random binding of IMP and NAD to the E·K^+ complex because at the ~0.1 M concentration of K^+ needed for saturation the second-order binding rate of K^+ is much faster than the binding rates for IMP or NAD at their typical ~10^{-4} M concentrations. During steady-state turnover a significant proportion of the enzyme is in complexes with the XMP product since XMP release is partially rate limiting. This mechanism was supported by the subsequent demonstration with IMPDH from other sources that substrate inhibition by NAD is uncompetitive with respect to IMP and reflects NAD binding to E·XMP or E-XMP* (53). Uncompetitive substrate inhibition is not uncommon in reactions with ordered product release (90, 91). Heyde and Morrison commented that if water forms a complex with the enzyme and is present at a saturating concentration, it must add before K^+ and NAD or different steady-state patterns would be seen (83, 84). The kinetic constants for the hydrolysis of 2-Cl-IMP by the human type 2 IMPDH are independent of both K^+ and NAD, consistent with the addition of water early in the pathway (30, 92). The kinetic data for the A. aerogenes enzyme suggested compulsory dissociation of M^+ during each turnover (83, 84). The observation that high concentrations of M^+ inhibit many IMPDH, and ions that activate at lower concentrations often inhibit at lower concentrations, may reflect that the M^+ must dissociate to regenerate the free enzyme in each catalytic cycle (16). However, equilibrium studies with the A. aerogenes IMPDH showed that IMP binding was not altered by K^+. Equilibrium binding studies with the human type 2 protein showed that IMP bound at a stoichiometry of four sites per tetramer and the affinity was only enhanced two-fold by K^+ (16). Significant NAD binding was not detected for either the *A. aerogenes* or human type 2 protein, but NADH was found to bind to the latter protein, and all substrates/products bind individually to the *T. foetus* protein (18).

A practical impediment in studies of the IMPDH mechanism has been that the reverse reaction is so slow that it is difficult to measure, even when the NAD that might form is removed by a coupling enzyme system. Heyde and Morrison were able to study isotope exchange at equilibrium by lowering the pH to 7.0 from the value of 8.1 used for the forward reaction (83). Heyde and Morrison reported that at pH 7.0 the equilibrium constant

$$K_{eq} = \frac{[NADH][XMP][H^+]}{[NAD][IMP]}$$

is 14.5×10^{-7} M. At pH 7.0 the mechanism was not rapid equilibrium, rather IMP and/or XMP dissociation were rate limiting. Whereas it is clear that H^+ is a product of the reaction, it has often been neglected that the purine ring of the product XMP ionizes to the anion at neutral pH, and thus the reverse reaction may depend on $[H^+]^2$.

The steady-state kinetic data for IMPDH from numerous organisms subsequently have been interpreted in terms of a predominant kinetic pathway with ordered addition of IMP before NAD and release of NADH before XMP. These studies generally relied upon product inhibition rather than including dead-end inhibition studies, which limits the view of the available pathways (93). The random pathway for the human and *T. foetus* enzymes was discerned from deuterium kinetic isotope effect (KIE) measurements (18, 87, 89). When the 2-hydrogen of IMP that is abstracted during the reaction was replaced by deuterium, a significant primary ^2H effect on k_{cat}/K_m ($^DV/K$) of both substrates was observed. The presence of a KIE on k_{cat}/K_m requires that the substrate dissociate at a rate that is at least comparable to that of the first irreversible isotope-sensitive step in the forward reaction (90, 91). Thus, the observation of ^2H KIE effects on k_{cat}/K_m for both substrates unambiguously demonstrated that both IMP and NAD can dissociate from the ternary E·IMP·NAD complex and that the first irreversible step of the reaction succeeds hydride transfer. The larger $^DV/K$ effect for NAD than IMP showed that NAD dissociates more rapidly from the ternary E·IMP·NAD complex. Apparently, the affinity of NAD for the unliganded enzyme is low enough that under normal experimental conditions IMP binds before NAD, thus providing the illusion of an ordered substrate addition mechanism. In no case was a ^{2}H KIE on k_{cat} observed, indicating that hydride transfer is not rate limiting in turnover.

A. CASE STUDIES

The sections below are grouped according to the organism from which the IMPDH (gene) was obtained since each of the studies has a peculiar character.

1. Tritrichomonas foetus IMPDH

A particularly thorough study of the *T. foetus* enzyme demonstrates the insight that can be obtained by combining a variety of methods (18). Stopped-flow studies took advantage of the tryptophan fluorescence of the *T. foetus* enzyme and revealed that IMP bound in two successive steps, the second of which might reflect reorganization of the protein to embrace the ligand (see Figure 8A). Since two-step IMP binding was also seen with a mutant in which the reactive cysteine was replaced by serine, the formation of the covalent adduct apparently occurs in a later step. The isomerization of the E·IMP complex is approximately threefold faster than dissociation and 75% of the rate of hydride transfer. The isomerization has an equilibrium constant

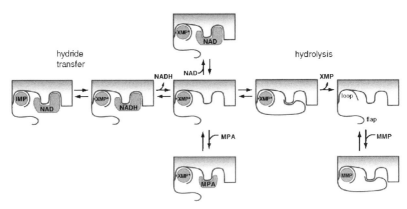

Figure 9.   Plausible structural dynamics of the IMPDH catalytic cycle. The active-site loop 6 and the flap appear to be relatively disordered in the absence of substrates. Binding of IMP causes a conformational change in the loop; the flap remains largely disordered. After the hydride transfer is complete, NADH dissociates and the flap moves into the NAD site, activating water and converting the enzyme to a hydrolase. (See insert for color representation.)

of 20, and reversal of this conformational change is rate limiting for IMP dissociation. In contrast, the binding of the product XMP did not show evidence of an isomerization, consistent with the failure of XMP to bring order to loop 6 in the E·XMP crystal structure (18). The rate of XMP release of $17 \, s^{-1}$ is eightfold greater than k_{cat}, showing that an earlier step is rate limiting.

NAD and NADH binding were also measured by changes in *T. foetus* IMPDH protein fluorescence (18). Interestingly the NAD binding curve was sigmoidal with a Hill coefficient of 2 suggesting cooperativity in binding. NAD analogs with altered reduction potentials (e.g., 3-acetylpyridine adenine dinucleotide, APAD, $E^{0'} = -0.258$ V, compared with NAD, $E^{0'} = -0.320$ V) gave similar k_{cat} values, implying a common rate-determining step because the intrinsic rates of hydride transfer should be modulated by the magnitude of the differences in reduction potential between the substrates and products. This common rate-determining step might be E-XMP* hydrolysis or subsequent XMP dissociation. In support of a late rate-limiting step, the KIE from [2-^{2}H]IMP was unity for k_{cat} with both NAD and APAD. A $^D V/K$ of 1.5 for IMP but a clearly larger value of 2.2 for the dinucleotide was found in the pre-steady state, verifying that the latter is not as "sticky" as IMP.

An example of different spectroscopic techniques illuminating distinct phenomena was found in pre-steady-state measurements monitoring absorption or fluorescence of the reduced nicotinamide chromaphore. Absorption changes showed that the KIE from [2-^{2}H]IMP on the burst of NAD formation was only 1.4, severalfold too small to be the intrinsic KIE, and thus nonchemical steps contribute to the rate. The amount of NADH formed in the burst is only 0.5 per subunit, suggesting the hydride transfer step is near equilibrium with an equilibrium constant of unity, a sign of a high-performance catalyst (94, 95). Consistent with this interpretation, the amount of product formed in the burst increased to 0.8 per subunit with the more readily reduced 3-acetylpyridine adenine dinucleotide. In contrast, fluorescence changes did not show isotope effects, and the observed rates were slower than those seen in the absorption studies. Apparently NADH fluorescence is quenched on the enzyme and the rate of fluorescence change reflects NADH dissociation. A step after NADH dissociation is rate limiting. Earlier studies had shown that a covalent enzyme–substrate (or product) adduct was detectable only in the presence of NAD (17). Rapid quench measurements of acid-precipitable radioactive material formed from [^{14}C]IMP revealed that during the steady-state turnover ~50% of the total enzyme was present as E-XMP* complexes (E-XMP*·NADH, E-XMP*) (18). The conglomerate results show that both NADH dissociation and E-XMP* hydrolysis are partially rate limiting.

The *T. foetus* Cys319Ser mutant studies showed the insight that can be obtained by careful multifaceted comparison of mutant and wild-type enzymes. In the mutant, the two-step binding process for IMP remains; a 40-fold decrease in IMP affinity primarily results from reduction in the forward rate of the isomerization step rather than changes in an initial binding step. Because the mutant is essentially inactive ($<6 \times 10^{-4}$ of wild type), rates of IMP binding to E·NAD and NAD binding to E·IMP could be measured. Perhaps surprisingly, the rates of IMP and NAD binding to the binary complexes were comparable to the rate of the bimolecular binding step to the free enzyme in both cases, suggesting unimpeded access to their binding sites rather than cloaking of one site by binding of the other substrate. Rates of dissociation from the ternary complex were 65-fold slower for IMP and approximately 1.5-fold faster for NAD compared to the respective binary complexes.

2. *Escherichia coli IMPDH*

The functional properties of the *E. coli* enzyme have also been characterized in detail, although a crystal structure has not been reported (21, 65). This enzyme has a 30-fold higher k_{cat} than the human enzyme and has measurable activity in the absence of a M^+ activator, making it a particularly good system for the study of slower reactions. The rate of product release limits turnover as in the case of other IMPDH. Hedstrom and co-workers (21, 65) point out that the higher catalytic ability compared to other IMPDHs results from increased proficiency in multiple steps, including hydride transfer, NADH release, and E-XMP* hydrolysis, an integrated optimization that one might expect from natural evolution (94, 95). Pre-steady-state studies of the NADH formation revealed that the rate of fluorescence change was smaller than that the rate of absorption increase, as seen with the *T. foetus* enzyme. The authors (21, 65) suggested that the fluorescence of bound NADH is quenched by stacking with the purine of XMP.

Mutants of the *E. coli* enzyme were constructed at each of 11 evolutionarily conserved aspartate and glutamate residues before any IMPDH crystal structures were available (65). The acidic residues at the IMP and NAD binding sites were both identified. The effects of alanine substitution on K_m, k_{cat}, and K^+ activation as well as inhibition by mizoribine-monophosphate and TAD were characterized (65). Six of the mutations had no significant functional effect. The Asp338Ala mutant (corresponding to human Asp364;

see Figure 10A) affected k_{cat} more than 600-fold and increased the K_m for IMP, consistent with the crystallographically observed binding of the homologous residue to the ribose hydroxyls of the substrate in other IMPDHs and suggesting a role in catalysis.

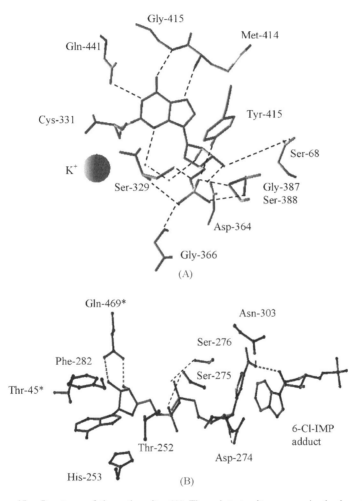

Figure 10. Structures of the active site. (A) The substrate site as seen in the hamster E-XMP*·MPA structure (28). (B) SAD bound to the NAD site as seen in the complex with the human type 2 6-Cl-IMP adduct (31). (C) The potassium binding site as observed in the hamster IMPDH. Residues marked by asterisks are from a second subunit (28). Figure made using the program DeepView (95a). (See insert for color representation.)

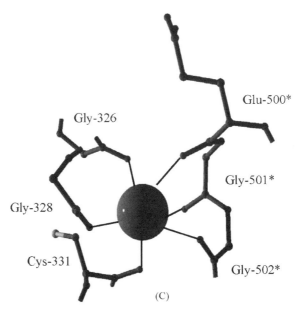

Figure 10. (*Continued*).

Because the k_{cat} of at least the wild-type enzyme is determined by product release, the magnitude of mutational effects on the catalytic steps may be masked in steady-state studies. Thus, the characterization of the Asp338Ala mutant via kinetic isotope effects from [2-^{2}H]IMP was particularly enlightening (96). Kinetic isotope effects were observed on both k_{cat} and k_{cat}/K_m(IMP) for the mutant, with values of 2.5 and 3.4, respectively, whereas no KIE were observed for the wild-type enzyme. Thus, upon mutation hydride transfer has become a rate-limiting step. Pre-steady-state experiments revealed that the burst of NADH formation observed with the wild-type enzyme is no longer seen in the mutant, and the hydride transfer rate is diminished at least 5000-fold. The rate of inactivation of the Asp338Ala mutant by 6-Cl-IMP is *increased* 3-fold, indicating that the nucleophilicity of the nucleophilic active-site sulfhydryl is relatively unperturbed; the k_{cat} for hydrolysis of 2-Cl-IMP decreased 8-fold from 3% of the IMPDH reaction rate with the wild-type enzyme. These results were considered to be consistent with the structure of the human IMPDH in which the Asp338 homolog hydrogen bonds to the 2-OH of IMP, which is also hydrogen bonded to the carboxamide of

SAD, linking the two substrates. In accord with this coupling, the wild-type enzyme-catalyzed rate of hydride transfer from 2′-deoxy-IMP decreased by >5-fold; furthermore, in the 2′-deoxy-IMP reaction both absorbance and fluorescence changes showed the same rates, suggesting that the nicotinamide and purine rings are no longer oriented to quench NADH fluorescence in the bound complex. The Asp248Ala substitution selectively impaired NAD binding, consistent with the crystallographically observed role of the corresponding residue in human IMPDH (Asp274) in binding the ribose hydroxyls of the selenazole riboside moiety in the SAD complex (31).

Mutations of three acidic residues were found to affect K^+ activation: Asp13, Asp50, and Glu469. The Asp13Ala (human Asp34) and Glu460Aala (human Glu500) proteins are activated by Mg^{2+} and Ca^{2+} in lieu of K^+, a very unusual feature for a cation-activated protein; the Asp50Ala (human Asp71) mutant was inhibited by these divalent cations (21). These results provided impetus for the proposal that more than one K^+ interacts with the protein.

3. Human IMPDH

Mass spectrometric analysis of a radiolabeled peptide formed during turnover of [8-^{14}C]IMP by the type 2 enzyme revealed that several percent of the enzyme was present in a covalent complex with properties consistent with a linkage between C-2 of IMP and cysteine 331, now known to be the oxidized intermediate denoted E-XMP* (97). The amount of E-XMP* increased more than 30-fold in the presence of MPA, consistent with MPA trapping the reaction intermediate. Fleming et al. showed that the formation of the MPA-inhibited complex from the product XMP was slow with a half time of 178 min (70). The observed rate demonstrates that there is a slow step in formation of the E-XMP*·MPA complex; this step could be either formation of the putative tetrahedral enzyme–XMP intermediate, its dehydration to E-XMP*, or an isomerization that leads to the final E-XMP*·MPA complex.

Kinetic isotope effect studies of the human type 2 enzyme revealed that the rate-limiting step is not altered by substrate deuteration or by D_2O and thus does not involve either hydride transfer from IMP or proton transfer (87, 89). However V/K values for both substrates were affected by deuteration of IMP and by D_2O, showing that both types of transfers occur in steps before the first irreversible step of the reaction. Wang and Hedstrom found that k_{cat} for the human type 2 enzyme was similar for a series of NAD analogs despite variations in their intrinsic reduction potential, also indicating that hydride transfer was not rate limiting (87). The net reverse reaction did not occur to a measurable extent, consistent with the studies of the *A. aerogenes* enzyme (83). However, NADH did reduce thio-NAD in the presence of IMP, demonstrating reversible hydride transfer. NADH could trap all of the E-XMP* formed in the reaction of thio-NAD and IMP showing that product release is necessarily ordered, with NADH dissociating before E-XMP* hydrolysis (if release was random some XMP would escape from E-XMP* rather that being trapped, even in the presence of a saturating NADH concentration).

B. LIGAND BINDING

1. IMP Binding Site

The tautomeric structure of IMP and XMP is of importance in discrimination of these nucleotides from physiologically abundant monophosphates such as AMP and GMP. Crystallographic studies indicate that IMP, XMP, and the covalent E-XMP* are bound as the 6-keto, N1–H forms, which are the predominant forms in solution (98, 99). XMP ionizes at N3 to give the 6-keto, 2-enolate form, with a pK of 5.7. It is unknown whether IMPDH catalyzes the ionization of the product.

The hamster IMPDH structure (PDB file 1jr1) revealed the covalent E-XMP* adduct with cysteine 331 and its environs, as illustrated in Figure 10A. The interacting residues are conserved in most IMPDH sequences. Many of the enzyme–ligand interactions arise from the protein backbone. The nucleoside in the various IMP, XMP, and E-XMP* crystal structures (Table 2) has an anti configuration about the glycosidic bond which leaves the 2-position accessible for reaction; the conformation is similar to that determined for IMP with the human type 2 enzyme using transferred-nuclear Overhauser effect (NOE) nuclear magnetic resonance (NMR) methods (88). This congruence has not always been obtained in studies of protein–nucleotide complexes by NMR and crystallographic methods (cf. ref. 100). The anti conformation is preferred by IMP in solution, and binding this conformer maximizes the association rate. The sugar conformation is variable among complexes between C2′ and C3′ endo or exo, in keeping with the relatively small energy differential among sugar-puckering conformers.

An extensive mutagenesis study of human type 2 IMPDH (101) focused on changes in k_{cat} and K_m. Mutations in the IMP binding site (Gly326Ala,

Asp364Ala, Cys331Ala) resulted in <0.1% activity. Gly326 is located in a turn in the base of the IMP site whereas the other two residues form more obvious contacts (Figure 10A). Mutations of other residues that partake in binding IMP (Met70Ala, Tyr111Ala, and Lys409Ala) yielded low-activity proteins with increased K_m values. A Ser329Ala mutation increased the K_m for both IMP and NAD without altering k_{cat}. Mutations of residues near the purine in E-XMP* (Thr333 whose side chain OH is hydrogen bonded to the side chain of Gln441) yielded 56 to 200-fold decreases in k_{cat} with a selective increase in the K_m of NAD. Thr333 and Gln441 were proposed to catalyze the hydrolysis of E-XMP* (101), although based on typical pK_a values no chemical mechanism is obvious.

In cases where the protein–IMP interactions involve the peptide backbone, the use of modified substrates can be more informative than protein mutagenesis. IMP derivatives lacking the ability to interact at the N1, N3, O6, or N7 positions were found to be inhibitors rather than substrates but had affinities comparable to IMP itself (30). The lack of substrate activity might be due to inability to delocalize negative charge that accrues at least transiently during bonding changes in the reaction, perturbations that are more important for transition state stabilization rather than initial recognition (30).

Remarkably, the purine nucleotide in the structure of the human type 2 isozyme with the covalent 6-Cl-IMP adduct (1b3o) occupies the same position as the nucleotide in the hamster structure, although the covalent linkage to cysteine 331 is at C6 rather than at C2 (31). Comparison of the human and hamster structures indicates that the core $(\beta/\alpha)_8$ barrel provides a relatively constant configuration whether or not the NAD or K^+ sites are occupied, whereas the connecting loops and flaps reorganize (Figure 6B). Thus, the plasticity of the protein loop residues 325–340 (mammalian numbering) is sufficient to mold around the altered substrate site. In the human structure, loop residues replace some of the flap contacts with the barrel seen in the hamster structure. The active-site flap residues 400–450 remain largely disordered. The structure of the human type 2 IMPDH complex with the IMP analog ribavirin monophosphate has been described in which the flap residues 420–437 reorient to occupy the NAD binding site (cited in ref. 80).

A unique feature of the *T. foetus* enzyme (1ak5) is the presence of Arg 382 at the binding site for the phosphoryl group in the E·XMP complex; this position is occupied by uncharged residues in the other enzymes. This novelty was suggested as a target site for design of selective inhibitors (33); however, subsequently the structure of the *T. foetus* enzyme complex with IMP and β-CH$_2$-TAD (1lrt) found this arginine side chain to point away from the active site, casting doubt on its relevance for ligand interactions (20). Structures of ribavirin–monophosphate complexes with the *T. foetus* enzyme have been described (1me7, 1me8); however, the basis for differences in affinity among species was not deduced from these structures (19).

2. NAD Binding Site

The conformation of the bound dinucleotide has been deduced from a combination of transferred NOE NMR measurements with NADH (102) and the crystal structures of SAD (31) and β-CH$_2$-TAD complexes (20). Since NAD/NADH only bind productively when the substrate site is occupied, the NMR studies investigated the NADH conformation with human type 2 enzyme in the presence of saturating amounts of XMP, whereas crystal structures used the same protein modified by 6-Cl-IMP for the SAD complex (1b3o), and the *T. foetus* subdomain deletion protein was used with IMP for the β-CH$_2$-TAD complex (1lrp). The dinucletotide binds in an extended conformation with both glycosidic torsional angles in the anti configuration. The NMR data for the nicotinamide ring showed proximity of the carboxamide NH$_2$ protons to the pro-R hydrogen at the N4 position, leaving the reactive pro-S proton accessible for transfer. The stereochemical course of the reaction catalyzed by various IMPDH yields hydride transfer to the B side of the nicotinamide ring, that is addition of the pro-*S* proton (89, 103, 104). This stereochemistry is consistent with the empirical finding that dehydrogenase reactions for which the equilibrium lies far toward NADH formation utilize the B side of the nicotinamide ring (105). The carboxamide oxygen is positioned syn to the reactive H, as expected from observations for other enzymes and model studies that show that the carboxamide orientation is related to the stereospecificity of hydride transfer (106). In SAD and TAD the anti configuration of selenazole (thiazole) riboside preserves the close intramolecular Se(S) contact with O4′ of the ribose, an electrostatic interaction that restricts the conformation of SAD in solution and provides an entropically favorable anchor point for binding to IMPDH (67). This configuration is not compatible with the conformation of NAD bound to other dehydrogenases and contributes to the relatively low affinity of TAD for those enzymes (107).

In the hamster enzyme structure, the aromatic ring of MPA lies over the purine ring of E·XMP* in a mimic of the position of the nicotinamide of NAD

(28). The hydroxyl group of MPA occupied a position that authors proposed to normally bind the substrate water molecule, implying that MPA is a bisubstrate inhibitor.

The human enzyme structure (1b3o), illustrated in Figure 10B, showed that the adenosine portion of the SAD binding site is composed of residues from two subunits, whereas intersubunit contacts at the adenosine moiety of TAD were not seen in the T. foetus complex. Likewise, in the T. foetus E·XMP·NAD structure, NAD does not contact residues from another monomer (36). The selenazole (thiazole) is stacked against the purine ring of the substrate (analog) in a position that if occupied by a nicotinamide would allow hydride transfer to occur with the established B-side stereochemistry. Differences between the NAD binding sites of the human type 1 and type 2 IMPDH occur near the adenosine moiety, which stacks between the imidazole of histidine 253 and the aromatic ring of phenylalanine 282 and also makes a contact with threonine 45. In the human type 1 enzyme, the residues at these positions are arginine, tyrosine, and isoleucine, respectively. The cognate residues in the T. foetus enzyme (1lrt) are tryptophan 269, arginine 241, and isoleucine 27, the last of which does not interact with the ligand. The adenosine portion of the dinucleotide appears to provide a target for development of compounds selective for one or another IMPDH (47).

In the human type 2 enzyme structure, a hydrogen bond is present between the 2'-hydroxyl of the IMP analog and the carboxamide of the nicotinamide, suggesting a role in positioning for catalysis. Although 2'-deoxy-IMP has 55% of the k_{cat} of IMP, the rate-limiting step for the analog reaction is not known, which diminishes the utility of the observation (30). Hence, the large decrease in the rate of hydride transfer with 2'-deoxy-IMP and the E. coli enzyme is probably more directly incisive. Alanine mutations created in the NAD site of the human type 2 enzyme have also been characterized (101).

IV. Chemical Mechanism

This section addresses catalysis of the covalent changes that occur in the reaction in the context of the structural, kinetic, and equilibrium data described above. A plausible mechanism is illustrated in Figure 8B. This scheme includes putative tetrahedral intermediates E-IMP and E-XMP and does not explicitly illustrate noncovalent (Michaelis) complexes. The aromaticity of the purine is disrupted in the tetrahedral species, and these might be high in energy compared to the conjugated E-XMP*. Key catalytic features include the stacking of the purine and nicotinamide rings to allow hydride transfer and acid–base catalysis of the formation and hydrolysis of the covalent enzyme intermediates.

Halo-purine nucleotides have been particularly informative probes of IMPDH. The pH dependence of the inactivation rate of the *E. coli* IMPDH at subsaturating 6-Cl-IMP concentration revealed a pK value of 8.4 for a group that must be deprotonated for inactivation; this group was attributed to the reactive sulfhydryl (62, 63). A similar study with the human enzyme at saturating 6-Cl-IMP gave a pK of 7.5 in the E·6-Cl-IMP complex (30). These pK_a values are in a typical range for cysteine thiols in peptides, and thus there does not appear to be an extraordinary activation of the nucleophile (108).

The hydrolysis of the 2-fluoro and 2-chloro analogs of IMP to XMP and the halide ion are reactions in which neither K_m nor V_{max} is affected by NAD or monovalent cation (30, 64, 92). Figure 8B illustrates how these reactions might fit into the sequence for the normal IMPDH reaction. The cation independence of these rates supports the notion that the monovalent cation is primarily involved in the NAD-dependent segment of the dehydrogenase reaction (30, 55, 61, 96). The k_{cat} values of the dehalogenation of 2-F- and 2-Cl-IMP by human type 2 IMPDH were approximately the same and only about fivefold slower than the k_{cat} for the IMPDH reaction. Thus, the rate-limiting step occurs before any step that is common to the IMPDH reaction (e.g., hydrolysis of the E-XMP* intermediate or XMP dissociation). The same rates of product formation from substrates with C–F and C–Cl bonds, which have substantially different intrinsic strengths, indicate that bond cleavage is unlikely to be rate limiting. D_2O did not alter the k_{cat} or k_{cat}/K_m for 2-Cl-IMP, indicating that proton transfers are relatively fast in this reaction (30). Strangely, the pH–rate profiles for the hydrolysis of 2-Cl-IMP did not reveal dependence on any group that needed to be deprotonated, only a group with a pK of 9.9 that must be protonated, which is similar to a pK seen in the normal IMPDH reaction (30). This observation suggests that the hydrolysis of 2-halopurines might not require formation of the covalent E-XMP* adduct but rather water directly displaces the halide from the more electrophilic C2 position. The possibility that the 2-halo-IMP reactions follow a distinct mechanism without covalent adduct with enzyme has yet to be resolved.

The pH–rate studies of the physiological IMPDH reaction with the human enzyme showed that both acidic and basic groups are required for maximal activity at both saturating and nonsaturating substrate levels. None of the pK values correspond to the free substrates, and thus they arise from the enzyme or enzyme–substrate complexes. The acidic group is attributed to the nucleophilic cysteine, whose pK is variously estimated as 7.2 from k_{cat} and 6.9 from k_{cat}/K_m (NAD). A basic group with a pK of 9.4 is seen for k_{cat}/K_m of NAD and 9.8 from k_{cat}. The molecular nature of this basic group is uncertain. The interpretation of the pH variations for the k_{cat}/K_m of IMP is complicated because IMP binding is not in rapid equilibrium; thus the observed pK values (8.1 and 7.3) could be perturbed from their intrinsic values; furthermore, because the pK values are separated by less than one unit, they are in fact not distinguishable (90). Thus, it is not clear whether the pK of the cysteine should be assigned to 8.1 in the free enzyme and decreases in the presence of substrates or is approximately constant with a value of ~ 7.2. In either case, the basic group that contributes to the early stages of the reaction, which include IMP binding, has a much lower pK than that seen later in the reaction. Again, the molecular assignment of this ionization is unclear. The D_2O kinetic isotope effects for the k_{cat}/K_m of both IMP and NAD [1.4 and 2.0, respectively (89)] show that proton transfers are kinetically important early in the sequence. The absence of a KIE from D_2O on the turnover rate (k_{cat}) suggests that k_{cat} is not limited by a step involving proton transfer and is thus unlikely to include deprotonation of water for hydrolysis of E-XMP*. In sum, the molecular constituents of the basic groups remain enigmatic.

V. Protein Conformational Flexibility and Mechanism of Water Activation

The relationship of protein dynamics to enzyme function is currently a topic of widespread interest (109). The structural, functional, and kinetic aspects of protein loop and domain movements have been discussed (38, 110–114, 116–119), including in our recent work (120). Movement of IMPDH protein segments is used both to sequester the intermediate and to recruit groups from the flap into the active site. The large movement observed for the subdomain reorientation as shown in Figure 6C plays an unknown role in protein function, one that is not necessarily related directly to IMPDH catalytic activity. Structural studies of many proteins have shown that rearrangements of loops that constitute active-site lids may be rigid-body reorientations around flexible hinges or local changes in the conformation of the polypeptide chain (119). In some cases the motion appears to be independent of the protein ligation state while in other instances the rates are modulated by the binding of ligands, but generalizations are not yet clear.

Crystal structures show substantial alterations among structural elements in different complexes, as illustrated in Figure 6. Perhaps most amazing is the movement of the loop containing the active-site cysteine by as much as 7 Å in enzyme complexes with 6-Cl-IMP (1jcn, 1b3o, 1nfb) so that the nucleotide in the E·IMP, E-XMP*, and 6-Cl-IMP modified proteins is in the same position although the linkage between the protein and the purine varies between C2 and C6 (Figure 6B).

The existence of different conformations observed in various crystalline complexes is supported by limited proteolysis data that showed changes in loop susceptibility in solution. A study of the potential conformational alterations associated with ligand binding to the human type 2 IMPDH revealed that residues in the flap region (from residues 411 to 441) became protected from various proteases in the presence of IMP or XMP (121). NADH provided less protection and NAD did not protect at all, in accord with other binding data (16). Interestingly, the 6-Cl-IMP inactivated enzyme was not protected from proteolysis, consistent with the alternative conformation of the flap seen in the crystal structure (PDB files 1jcn, 1b3o, 1nfb; see also Figure 6B). The conformational study also found no significant alteration in the far-ultraviolet (UV) circular dichroism spectrum in the presence of ligands, indicating that major rearrangements of secondary structure did not occur.

A calorimetric study of human IMPDH reported that the first three IMP bound with the same affinity but the fourth IMP bound up to eightfold more weakly, reflecting negative cooperativity (122). The ΔG values for successive IMP reflected different combinations of ΔH and ΔS, indicating that at least subtle differences were present in the physical binding processes and were masked through enthalpy–entropy compensation. Negative cooperativity for IMP has not been reported in kinetic or other binding studies, perhaps in part due to the ranges of substrate concentrations employed. The independence of the binding enthalpy on the type of buffer indicated that no protons were released in conjunction with complex formation, suggesting that the negative cooperativity did not arise from changes in the charge of the system. IMP binding showed a large and temperature-dependent heat capacity change [-1.8 kcal/mol (IMP bound) at 25°C] indicative of a conformational change on binding (123, 124). The authors proffered an explanation that the ligand-free enzyme exists in multiple conformational forms, which is consistent with the difficulties in crystallizing the ligand-free protein. The calorimetric results for MPA binding to the E·IMP complex were evaluated in terms of a protein conformational distribution that persists in the binary complex.

An alignment of the *B. burgdorferi* apo enzyme and *T. foetus* IMP-bound structures allowed Prosise and Luecke to suggest a structural explanation for the random-in ordered-out kinetic mechanism (36). When IMP binds first, the enzyme is in the open conformation and IMP has a free access to the active site. When NAD binds first, it blocks one entrance to the active site and IMP must use another route, which is created by a 10 × 12-Å opening formed by the movement of the loop and the flap. When IMP binds, the active-site "lid" closes and XMP has to wait until NADH dissociates before XMP itself can leave the active site.

A series of structural and kinetic studies of *T. foetus* IMPDH conducted in the Hedstrom laboratory have led to a new model of the intimate association between IMPDH conformational flexibility and catalysis. In contrast to the previously published IMPDH structures, the entire flap appears to be ordered in the *T. foetus* E·MMP complex (35), which allowed several new observations. The distal portion of the flap interacts with the NAD binding site, with the conserved Arg418–Tyr419 dyad occupying the nicotinamide subsite (see Figure 11). This flap orientation creates a previously unobserved closed active-site conformation. Residues 416–419 are organized in an α-helix,

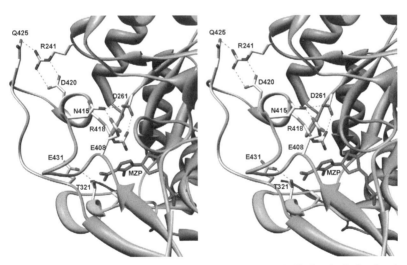

Figure 11.   Stereoview of the interactions of the flap with the NAD binding site in the closed enzyme conformation (*T. foetus* E·MMP complex; PDB file 1pvn). MZP is shown in magenta sticks. The distal flap (residues 413–431) is orange, the active-site loop 6 (residues 313–328) with the active-site Cys is cyan. Hydrogen bonds are shown as dotted lines. (See insert for color representation.)

suggesting that the flap maintains its helical structure while it moves in and out of the NAD site. In addition to the hydrogen-bonding pattern characteristic of IMP binding, several new hydrogen bonds were identified in the *T. foetus* E·MMP structure, which may resemble the E-XMP* transition state. The conserved Arg418–Tyr419 dyad, located in the flap, forms hydrogen bonds with a water molecule which is poised to attack C2 of bound XMP*, consistent with a role of the dyad in water activation in the hydrolysis step. Thus, movement of the flap appears necessary for water activation and has been postulated to convert the enzyme from the open (dehydrogenase) conformation to the closed (hydrolase) conformation. Substrate inhibition by NAD results from formation of a dead-end E-XMP*·NAD complex, preventing the enzyme from assuming the hydrolase conformation (35, 125).

Important mechanistic insights were obtained by characterization of *T. foetus* Arg418 substitution variants. Substitution of Arg418 with Ala decreased steady-state enzymic activity (k_{cat}) 500-fold but the presteady burst of NADH production in the first enzyme turnover was unaffected, nor was the rate of NADH release (35, 125). Thus, Arg418 substitution does not affect the hydride transfer step and specifically impairs the E-XMP* hydrolysis. A similar (though much less pronounced) result was obtained for a Tyr419Phe substitution (125), but the lack of change in pH dependence of the hydrolysis reaction in this mutant suggested that Arg418, and not Tyr419, is directly implicated in water activation (126). Arg418 and Tyr419 also stabilize the closed flap conformation (125). Arg418 forms a hydrogen bond to Asp261 of the dinucleotide site (35). A substitution of Arg418 with Lys did not impair E-XMP* hydrolysis but it destabilized the closed conformation, resulting in a strong enzyme inhibition by NAD and NADH. In contrast, an Arg418Gln substitution stabilized the closed conformation but the mutant was defective in E-XMP* hydrolysis, suggesting that the closed conformation per se is not sufficient for hydrolysis (126). Thus, Arg418 is the only residue that is directly implicated in water activation for the hydrolysis step of IMPDH reaction, and the other functionally significant flap residues mainly participate in flap stabilization and movement. The resulting model of IMPDH conformational transitions is presented in Figure 9.

Kinetic experiments indicate that deprotonation of Arg418 both stabilizes the closed conformation and creates a base that activates water for hydrolysis. The pK_a of Arg418 is ~8 in the closed conformation and ≥10 in the open conformation. However, Arg residues usually have pK_a values of ~13 and it remains unclear what perturbs the pK_a of Arg418 in the IMPDH structure (126). Interestingly, guanidine derivatives have been shown to rescue the Arg418Ala mutation of *T. foetus* enzyme (127). This effect has been attributed to an acceleration of the E-XMP* hydrolysis and not to a stabilization of the closed conformation. A solvent D_2O effect is observed at low guanidine concentrations, indicating that proton abstraction from water is rate limiting (127).

VI. Species-Specific Drug Selectivity

The origin of the species-selective affinity of MPA (see Table 1) has been a subject of several investigations (86, 128, 129). The *T. foetus* IMPDH has a 450-fold lower affinity for MPA than does the human enzyme but differs in only two amino acids in the MPA binding site. This corresponds to the changes in the Arg322Lys and Gln441Glu (human sequence numbering; the corresponding *T. foetus* residues are 310 and 431, respectively). The introduction of the arginine and glutamate in the framework of the *T. foetus* enzyme enhanced the affinity of MPA by 20-fold. A multiple inhibition study of the wild-type human type 2 and *T. foetus* enzymes, and a *T. foetus* (Lys310Arg)(Glu431Gln) mutant, used a combination of components of TAD, that is tiazofurin and ADP. The results showed the presence of substantial binding synergy for the *T. foetus* enzyme but not for the human enzyme. Apparently, the *T. foetus* enzyme uses some of the energy available from binding of ligands at the nicotinamide site (i.e., MPA or tiazofurin) to modify the protein conformation and thus enhance affinity at the adenosine site (129). The detailed nature of this conformational alteration had eluded crystallographic determination even in light of the structure of the complex with IMP and β-CH_2-TAD (20) until the recent crystal structure of *T. foetus* E·MMP complex indicated that the flap folds into the NAD site and makes contact with the same residues that bind NAD and MPA, as shown in Figure 11 (35). MPA and the flap thus compete for the nicotinamide subsite. The binding of MPA stabilizes the open-flap conformation, enhancing ADP affinity at the adenosine subsite.

Random mutagenesis was used to search for MPA-resistant forms of the human enzyme (128). Four mutations lying both near to and far from the active site were found. The authors noted that because the MPA inhibits by trapping of the E-XMP* intermediate, any mutation that decreases the accumulation of E-XMP* can result in reduced inhibition without directly modifying the MPA binding site (128). Indeed, evolution of IMPDH toward resistance to MPA may either change the MPA binding site topology or stabilize the closed enzyme conformation which will prevent MPA binding. Since MPA induces the open conformation and MMP induces the closed conformation, evolution of the enzyme toward resistance to these compounds should proceed in opposite directions. Indeed, an MPA-resistant Ala251Thr substitution variant of *Candida albicans* IMPDH is 4-fold less sensitive to MPA but is 40-fold more sensitive to MMP (130). The Ala251Thr substitution did not affect k_{cat} but decreased K_m values for both substrates, making the mutant enzyme more catalytically efficient. The mutation also rendered the enzyme resistant to NAD substrate inhibition. Therefore, the A251T substitution seems to stabilize the closed conformation, which has opposing effects on enzyme susceptibilities to MPA and MMP (130).

Saccharomyces cerevisiae encodes three functional IMPDHs: *IMD2*, *IMD3*, and *IMD4*. Any one of these genes supports growth in the absence of guanine, but only *IMD2* imparts MPA resistance, despite >90% sequence identity (131). This resistance phenotype is both due to an intrinsic resistance of IMD2 to MPA inhibition and its MPA-dependent transcriptional induction (132, 133). Generation of a set of chimeric *IMD2–IMD3* genes allowed identification of Ala253Ser as a critical amino acid substitution that is responsible for a large fraction of MPA resistance in IMD2 (134). A change in the electrophoretic mobility of *S. serevisiae* IMPDH was detected following MPA binding; this property allowed identification of the in vivo MPA-inhibited enzyme (133). The competition between MPA and the flap for the NAD site governs both MPA and MMP sensitivity and the rate of E-XMP* hydrolysis. In the human enzyme the open (dehydrogenase) conformation is favored, which results in MMP resistance, MPA sensitivity, and low k_{cat}. Bacterial enzymes favor the closed enzyme conformation, resulting in MPA resistance, MMP sensitivity, and fast turnover (2, 35).

VII. Monovalent Cation Activation

The requirement of IMP dehydrogenases for activation by monovalent cations such as K^+ was recognized by Magasanik (14) and recently reviewed (135). IMPDH from each species studied has been found to have maximal activity in the presence of millimolar concentration of K^+, the presumed physiological activator (16, 20–23, 83, 84, 136–138). However, both the tolerance for K^+ surrogates and the M^+ free activities of the enzymes vary considerably. The basal activity of the *E. coli* enzyme is ∼5% of its maximal value whereas the human type 2 enzyme has <1% of the maximal activity in the absence of added M^+. The millimolar affinities for M^+ have precluded

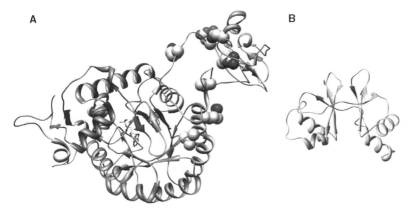

Figure 12. Subdomain (Bateman domain) of IMPDH. (A) Structure of human IMPDH type 1 (PDB code 1jcn) in complex with 6-Cl-IMP. The RP10-associated amino acid substitutions are shown in space-filling representation. The substrate is shown as sticks. (B) Structure of the Bateman domain of *S. pyogenes* IMPDH (PDB code 1zfj) (IMPDH residues 96–221). UCSF Chimera was used for structure visualization (36a). (See insert for color representation.)

direct determination of the number of cation binding sites in solution. The significance of this limitation has become evident in light of kinetic and crystallographic results that support multiple interactions (20, 21).

The K^+ site in the crystal structure of a hamster E-XMP*·MPA·K^+ complex described by Sintchak et al. is shown in Figure 10C (1jr1; (28)). The density ascribed to K^+ at this site 1 is surrounded by the main-chain carbonyl oxygens of Gly326, Gly328, Cys331, Glu500*, Gly501*, and Gly502* (where the * denotes a residue from a second subunit). Octahedral coordination by oxygen atoms is common for K^+ ions (139). The K^+ is ~6 Å from the C2 of XMP* and ≥8 Å from MPA. Sintchak et al. proposed that the hydroxyl of mycophenolic acid occupies the site of the nucleophilic water, thus implying that direct $M^+ \cdot OH_2$ coordination is not involved in the reaction. This is in accord with the M^+-independent rate of hydrolysis of 2-Cl-IMP. The structure of the ribavirin monophosphate complex of the *T. foetus* enzyme (1me8) showed a density attributed to a Na^+ ion at a similar location, coordinated to the carbonyls of Gly314, Gly316, and Cys319 from one subunit and Glu485*, Gly486*, and Gly487* from another subunit (19). Prosise et al. (19) proposed that in the presence of IMP the formation of the covalent intermediate is required for "recruitment" of the cation, but the smaller ring system in ribavirin monophosphate allows the cation to bind in the absence of a covalent linkage. Two groups have suggested that density ascribed to a water molecule in the *S. pyogenes* structure (water #179 in 1zfj) might actually be a cation, possibly NH_4^+ from crystallization buffer (19, 20).

The apparently conserved role of a cation at site 1 may be stabilization of a particular loop 6 conformation related to hydride transfer, rather than direct participation in the reaction. The absence of a cation in the human complex with 6-Cl-IMP-modified enzyme may be due to reorientation of the loop, which would malform the cation binding site; however, the cofactor site is evidently preserved since the 6-Cl-IMP-inactivated enzyme binds the NAD analog SAD.

A different K^+ binding site (site 2) was found near β-CH_2-TAD in the crystal structure of the *T. foetus* enzyme ternary complex with IMP (20). This site is at the dimer interface and comprises ligands from the side-chain hydroxyl of serine 22, both oxygens of the carboxylate of aspartate 264*, and the main-chain carbonyls of glycine 20, asparagine 460, and phenylanine 266*. Ion binding at this site could modify NAD binding or reactivity. The residues that provide side chains as ligands are not conserved among IMPDH; thus this site appears unlikely to be the one responsible for the unifying phenomenon of K^+ activation.

Two K^+ binding sites were observed in the subunit interface of *T. foetus* E·MMP structure (35). Site 1 is analogous to the one observed in the hamster IMPDH structure and site 2 is identical to the *T. foetus* E·β-CH_2-TAD K^+ binding site.

A kinetic study of the *E. coli* enzyme reported competitive inhibition with respect to both K^+ and NAD by Li^+, Na^+, Mg^{2+}, and Ca^{2+}, supporting interaction between the M^+ site and NAD site (21). In an Asp50Ala mutant, Mg^{2+} inhibition became noncompetitive with respect to K^+ and competitive with both IMP and NAD; in contrast to the wild-type enzyme, this mutant was inactive in the absence of K^+ (21). The change in inhibition pattern might be due to conformational alterations in the mutant or alternatively to the presence of another cation binding site. The presence of a second site was proposed by Kerr et al. since the residue analogous to aspartate 50 in the hamster enzyme lies >15 Å from the K^+ in the crystal structure (21). M^+ activation of the Glu469Ala mutant was positively cooperative, which also suggests more than one mode of interaction.

The comparable kinetic isotope effects from [2-^{2}H]IMP with Na^+, K^+, NH_4^+ and Rb^+ as activators of the human type 2 IMPDH (1.9- to 2.4-fold on V_{max}/K_m of IMP, 2.7- to 3.5-fold for V_{max}/K_m of NAD) suggest that the cations do not directly participate in the hydride transfer step of the reaction.

However, until studies ensure that the same steps in the mechanism are rate limiting in the reactions with the various ions, the isotope effects provide limited insight into the mechanism of M^+ activation (89). The K^+ independence of the pK for 6-Cl-IMP inactivation of the human type 2 IMPDH indicates that the M^+ does not dramatically alter the acidity of the cysteine–SH nucleophile, which is surprising given the coordination of the K^+ at site 1 to the carbonyl of this cysteine. The comparable pK values for the group that must be protonated for both the IMPDH reaction and the 2-Cl-IMP hydrolysis (9.4–9.9) suggest that the M^+ does not determine the ionization of this moiety. There are no indications that the M^+ has a role in the acid–base chemistry of IMPDH-catalyzed reactions. It appears that a role of the M^+ in activating the nucleophilic water has been excluded.

The theme emerges that M^+ accelerates the hydride transfer step of the reaction with apparently minor, if any, roles in binding of the substrate, determining the nucleophilicity of the reactive cysteine sulfhydryl or in the hydrolysis of the covalent E-XMP* intermediate. Perhaps the M^+ organizes the active site to place NAD in a particularly reactive conformation. M^+ activation by adjustment of protein conformation has been described for S-adenosylmethionine synthetase, dialkylglycine decarboxylase, pyruvate kinase, tryptophan synthase, and tyrosine phenol lyase and may be a common mode of action (140–144).

VIII.   Subdomain

The subdomain (Bateman domain) of IMPDH has no assigned in vivo function despite its nearly ubiquitous presence in the enzyme structure. The subdomain consists of ~120 residues that are inserted between helix α_2 and sheet β_3 and is unique to IMPDH among proteins with the $(\beta/\alpha)_8$ barrel fold. The topology of the subdomain is best defined in the S. pyogenes structure (23) and is shown in Figure 12B. The dimensions are ~ $20 \times 20 \times 30$ Å and it protrudes from the edges of the tetramer. In the structure of the human type 2 protein, the domain is rotated around hinges at residues 111–113 and 225–227 by nearly 120° relative to the orientation in the hamster structure (see Figure 6C). The functional significance of the rotation is unclear. The subdomain has two repeats of a sequence motif known as a "CBS domain" from its initial recognition in the amino acid metabolic enzyme cystathionine β-synthase (145). CBS domains have been found in a wide variety of proteins of apparently unrelated biological function, including protein kinases, ion channels, and enzymes, and generally occur in tandem pairs (for a review see ref. 146). A pair of CBS domains is also referred to as a Bateman domain. Each CBS motif shows approximate twofold structural symmetry in a sheet–helix–sheet–sheet–helix secondary structure (see www.sanger.ac.uk/Users/agb/CBS/CBS.html). Several structures of Bateman domains from different organisms and proteins have been published to date (23, 115, 147–150).

The apparent irrelevance of the subdomain to IMPDH enzymic activity was illustrated by a deletion that resulted in a direct connection of residues 100 and 227 in the *T. foetus* protein, a mutilation which caused only a 30% decrease in k_{cat} and no K_m alterations; the crystal structure showed that the core structure of the protein was unaffected (20). Similarly, replacement of the flanking domain with a tetrapeptide linker did not impair the in vitro catalytic activity of human type 2 IMPDH (151). An unequivocal demonstration of the physiological importance of the subdomain occurred in the year 2002 when two studies reported human patient families with autosomal dominant retinitis pigmentosa RP10 (adRP) that had alterations in residues within the subdomain of the type 1 isozyme (152, 153). These positions are marked by ** in Figure 5 and their spatial location on the IMPDH type 1 structure is illustrated in Figure 12A. AdRP is a genetically heterogeneous disease (see http://www.sph.uth.tmc.edu/RetNet; see ref. 154 for a review) resulting from an apoptotic loss of photoreceptors. The pathogenesis of IMPDH-associated adRP is not understood, but it is unlikely to result from a loss of IMPDH function since mice that are homozygous for IMPDH type 1 deletion develop only a mild retinopathy (155). Kennan and co-workers reported an Arg224Pro mutation (152), and an Asp226Asn change was discovered by Bowne et al. (153). The second group also found a Val268Ile change in another family but its relationship to the RP10 form of retinitis pigmentosa was less clear (153). Asp226 is conserved among all IMPDH proteins while Arg224 shows conservation among eukaryotic enzymes (see Figure 5). These amino acid substitutions did not alter IMPDH activity in vitro. Although a decreased protein solubility associated with adRP mutations was initially reported (155), another study concluded that adRP mutations are unlikely to affect protein folding and stability (156). Subsequent studies (157, 158) identified additional IMPDH type 1 amino acid substitutions, located in both the CBS subdomain and the core domain, that are associated with adRP and Leber congenital amaurosis: Arg231Pro, His372Pro, Thr116Met, Arg105Trp, Asn198Lys, and His296Arg (see Figure 12A). In agreement with previous findings, no mutation-associated changes in in vitro IMPDH activity were identified (157). Thus, the preponderance of evidence indicates the existence of a physiological function of the subdomain that may not be closely tied to IMPDH enzymic activity.

Hedstrom and co-workers demonstrated that IMPDH proteins from *E. coli*, *T. foetus*, and both human isoforms bind single-stranded nucleic acids with nanomolar affinity (159). This binding was sequence nonspecific and did not inhibit the dehydrogenase activity, suggesting that nucleic acids do not interact with the active site. A subdomain deletion variant of the *T. foetus* enzyme had increased K_d for nucleic acids, indicating that nucleic acids bind to IMPDH via interaction with the subdomain (159). Interestingly, the adRP-linked D226N and R224P mutations rendered IMPDH defective in nucleic acid binding in vitro and in vivo, which suggested that binding of single-stranded nucleic acids may be used as a functional assay for the adRP pathogenicity of IMPDH type 1 mutations (156). A physiological role for the nucleic acid binding by the IMPDH subdomain has not been demonstrated. Scott et al. postulated allosteric regulation of IMPDH by ATP binding to the subdomain (160). Addition of ATP induced a fourfold increase in the wild-type human IMPDH activity, and this effect was not seen in the adRP-linked IMPDH^{R224P} mutant. However, others have been unable to reproduce this work (156), and the literature suggests that it is not correct (85). A phosphorylation site of unclear function has also been identified in the subdomain of human IMPDH type 1 (161). Bacterial expression of the subdomain of the human enzyme yielded a folded protein as judged by circular dichroism (CD) spectra, which may exist as a monomer or dimer (151). No function was deduced from these studies, although an enhanced tendency of the subdomain-free core domain to form insoluble aggregates was noted (151).

The widespread occurrence of Bateman domains in a variety of unrelated proteins raises the question of whether a single role is fulfilled by all Bateman domains or if it is a common structural scaffold for a variety of functions. For instance, AMP-dependent protein kinase (AMPK) senses the energy–charge ratio by responding to the cellular ATP and AMP concentrations (160, 162, 163). A recent crystallographic study demonstrated that ATP and AMP alternate in binding to a Bateman domain located in the AMPK γ subunit (115, 148), thereby regulating kinase activity. Similarly, a crystal structure of ClC-5 chloride channel provided evidence for ATP and ADP binding by the CBS pair, but the physiological significance of this interaction remains unclear (149). In contrast, the Bateman domain of cystathionine β-synthase is involved in enzyme activation by *S*-adenosylmethionine (164) while the Bateman domain of the OpuA transporter from *Lactococcus lactis* has been postulated to act as a sensor for intracellular ionic strength (165). In addition, there is accumulating evidence that at least some Bateman domains may act as dimerization interfaces (150, 165, 166). It thus seems that the functions and binding partners of the Bateman domains may vary considerably between different proteins.

IX. Conclusions

The past five decades of work have provided a wealth of structural and mechanistic information on IMPDH and the parallel development of inhibitors of demonstrated clinical utility. In the last five years, a new concept of the enzyme structural dynamics in the catalytic cycle has been developed which incorporates midcycle enzyme transition from the dehydrogenase to the hydrolase conformation and water activation. Substantial light has been shed on the structural determinants of the species-specific drug selectivity and enzyme catalytic efficiency. Nevertheless, the substantive details of *how* the enzyme does its job remain unclear. The rates of the movements of protein loops and flaps in the reaction are largely unknown and their structures at various stages of turnover are far from completely defined. The chemical mechanism of catalysis still lacks definition of how the substrate water is activated and what perturbes the pK_a of the water-activating base residue. Furthermore, the exact mechanism of metal ion activation of IMPDH remains unknown. Perhaps most strikingly, the physiological role of an entire conserved domain, itself the size of some small enzymes, remains cryptic. IMPDH continues to be a fruitful ground for mechanistic and physiological research.

Acknowledgments

We would like to thank our colleagues whose work on IMPDH is cited in the references and to apologize to the multitude of authors whose contributions have not been cited due to the focus of the chapter or oversight. This work was supported by the National Institutes of Health and by an appropriation from the Commonwealth of Pennsylvania.

References

1. Buchanan, J. M., and Hartman, S. C. (1959) Enzymic reactions in the synthesis of purines, *Adv. Enzymol. 21*, 199.
2. Hedstrom, L., and Gan, L. (2006) IMP dehydrogenase: Structural schizophrenia and an unusual base, *Curr. Opin. Chem. Biol. 10*, 520.

3. Elion, G. B. (1989) The purine path to chemotherapy, *Science 244*, 41.

4. Pankiewicz, K. W., and Goldstein, B. M. (2003) *Inosine Monophosphate Dehydrogenase. A Major Therapeutic Target*, Oxford University Press, Washington, DC.

5. Weber, G. (1983) Enzymes of purine metabolism in cancer, *Clin. Biochem. 16*, 57.

6. Gilbert, H. J., Lowe, C. R., and Drabble, W. T. (1979) Inosine 5'-monophosphate dehydrogenase of Escherichia coli. Purification by affinity chromatography, subunit structure and inhibition by guanosine 5'-monophosphate, *Biochem. J. 183*, 481.

7. Gu, J. J., Stegmann, S., Gathy, K., Murray, R., Laliberte, J., Ayscue, L., and Mitchell, B. S. (2000) Inhibition of T lymphocyte activation in mice heterozygous for loss of the IMPDH II gene, *J. Clin. Invest. 106*, 599.

8. Zimmermann, A. G., Gu, J. J., Laliberte, J., and Mitchell, B. S. (1998) Inosine-5'-monophosphate dehydrogenase: Regulation of expression and role in cellular proliferation and T lymphocyte activation, *Prog. Nucl. Acid Res. Mol. Bio. 61*, 181.

9. Liu, Y., Bohn, S. A., and Sherley, J. L. (1998) Inosine-5'-monophosphate dehydrogenase is a rate-determining factor for p53-dependent growth regulation, *Mol. Biol. Cell. 9*, 15.

10. Sherley, J. L. (1991) Guanine nucleotide biosynthesis is regulated by the cellular p53 concentration, *J. Biol. Chem. 266*, 24815.

11. Jackson, R. C., Weber, G., and Morris, H. P. (1975) IMP dehydrogenase, an enzyme linked with proliferation and malignancy, *Nature 256*, 331.

12. Zimmermann, A. G., Wright, K. L., Ting, J. P., and Mitchell, B. S. (1997) Regulation of inosine-5'-monophosphate dehydrogenase type II gene expression in human T cells. Role for a novel 5' palindromic octamer sequence, *J. Biol. Chem. 272*, 22913.

13. Sirover, M. A. (1999) New insights into an old protein: The functional diversity of mammalian glyceraldehyde-3-phosphate dehydrogenase, *Biochim. Biophys. Acta 1432*, 159.

14. Magasanik, B., Moyed, H. S., and Gehrig, L. B. (1957) Enzymes essential for the biosynthesis of nucleic acid guanine; Inosine 5'-phosphate dehydrogenase of Aerobacter aerogenes, *J. Biol. Chem. 226*, 339.

15. Hager, P. W., Collart, F. R., Huberman, E., and Mitchell, B. S. (1995) Recombinant human inosine monophosphate dehydrogenase type I and type II proteins. Purification and characterization of inhibitor binding, *Biochem. Pharmacol. 49*, 1323.

16. Xiang, B., Taylor, J. C., and Markham, G. D. (1996) Monovalent cation activation and kinetic mechanism of inosine 5'-monophosphate dehydrogenase, *J. Biol. Chem. 271*, 1435.

17. Huete-Perez, J. A., Wu, J. C., Whitby, F. G., and Wang, C. C. (1995) Identification of the IMP binding site in the IMP dehydrogenase from Tritrichomonas foetus, *Biochemistry 34*, 13889.

18. Digits, J. A., and Hedstrom, L. (1999) Kinetic mechanism of Tritrichomonas foetus inosine 5'-monophosphate dehydrogenase, *Biochemistry 38*, 2295.

19. Prosise, G. L., Wu, J. Z., and Luecke, H. (2002) Crystal structure of Tritrichomonas foetus inosine monophosphate dehydrogenase in complex with the inhibitor ribavirin monophosphate reveals a catalysis-dependent ion-binding site, *J. Biol. Chem. 277*, 50654.

20. Gan, L., Petsko, G. A., and Hedstrom, L. (2002) Crystal structure of a ternary complex of Tritrichomonas foetus inosine 5'-monophosphate dehydrogenase: NAD+ orients the active site loop for catalysis, *Biochemistry 41*, 13309.

21. Kerr, K. M., Cahoon, M., Bosco, D. A., and Hedstrom, L. (2000) Monovalent cation activation in Escherichia coli inosine 5'-monophosphate dehydrogenase, *Arch. Biochem. Biophys. 375*, 131.

22. Zhou, X., Cahoon, M., Rosa, P., and Hedstrom, L. (1997) Expression, purification, and characterization of inosine 5'-monophosphate dehydrogenase from Borrelia burgdorferi, *J. Biol. Chem. 272*, 21977.

23. Zhang, R., Evans, G., Rotella, F. J., Westbrook, E. M., Beno, D., Huberman, E., et al. (1999) Characteristics and crystal structure of bacterial inosine-5'-monophosphate dehydrogenase, *Biochemistry 38*, 4691.

24. Umejiego, N. N., Li, C., Riera, T., Hedstrom, L., and Striepen, B. (2004) Cryptosporidium parvum IMP dehydrogenase: Identification of functional, structural, and dynamic properties that can be exploited for drug design, *J. Biol. Chem. 279*, 40320.

25. Robins, R. K. (1982) Nucleoside and nucleotide inhibitors of inosine monophosphate (IMP) dehydrogenase as potential antitumor inhibitors, *Nucleosides Nucleotides 1*, 35.

26. Ishikawa, H. (1999) Mizoribine and mycophenolate mofetil, *Curr. Med. Chem. 6*, 575.

27. Jayaram, H. N., Dion, R. L., Glazer, R. I., Johns, D. G., Robins, R. K., Srivastava, P. C., and Cooney, D. A. (1982) Initial studies on the mechanism of action of a new oncolytic thiazole nucleoside, 2-beta-D-ribofuranosylthiazole-4-carboxamide (NSC 286193), *Biochem. Pharmacol. 31*, 2371.

28. Sintchak, M. D., Fleming, M. A., Futer, O., Raybuck, S. A., Chambers, S. P., Caron, P. R., et al. (1996) Structure and mechanism of inosine monophosphate dehydrogenase in complex with the immunosuppressant mycophenolic acid, *Cell 85*, 921.

29. Collart, F. R., Osipiuk, J., Trent, J., Olsen, G. J., and Huberman, E. (1996) Cloning, characterization and sequence comparison of the gene coding for IMP dehydrogenase from Pyrococcus furiosus, *Gene 174*, 209.

30. Markham, G. D., Bock, C. L., and Schalk-Hihi, C. (1999) Acid-base catalysis in the chemical mechanism of inosine monophosphate dehydrogenase, *Biochemistry 38*, 4433.

31. Colby, T. D., Vanderveen, K., Strickler, M. D., Markham, G. D., and Goldstein, B. M. (1999) Crystal structure of human type II inosine monophosphate dehydrogenase: Implications for ligand binding and drug design, *Proc. Natl. Acad. Sci. USA 96*, 3531.

32. Risal, D., Strickler, M. D., and Goldstein, B. M. (2002) Binary complex of human type-I inosine monophosphate dehydrogenase with 6-Cl-IMP (PDB code 1JCN).

33. Whitby, F. G., Luecke, H., Kuhn, P., Somoza, J. R., Huete-Perez, J. A., Phillips, J. D., et al. (1997) Crystal structure of Tritrichomonas foetus inosine-5'-monophosphate dehydrogenase and the enzyme-product complex, *Biochemistry 36*, 10666.

34. McMillan, F. M., Cahoon, M., White, A., Hedstrom, L., Petsko, G. A., and Ringe, D. (2000) Crystal structure at 2.4 A resolution of Borrelia burgdorferi inosine 5'-monophosphate dehydrogenase: Evidence of a substrate-induced hinged-lid motion by loop 6, *Biochemistry 39*, 4533.

35. Gan, L., Seyedsayamdost, M. R., Shuto, S., Matsuda, A., Petsko, G. A., and Hedstrom, L. (2003) The immunosuppressive agent mizoribine monophosphate forms a transition state analogue complex with inosine monophosphate dehydrogenase, *Biochemistry* 42, 857.

36. Prosise, G. L., and Luecke, H. (2003) Crystal structures of Tritrichomonas foetus inosine monophosphate dehydrogenase in complex with substrate, cofactor and analogs: A structural basis for the random-in ordered-out kinetic mechanism, *J. Mol. Biol. 326*, 517.

36a. Pettersen, E. F., Goddard, T. D., Huang, C. C., Couch, G. S., Greenblatt, D. M., Meng, E. C., and Ferrin, T. E. (2004) UCSF Chimera — A visualization system for exploratory research and analysis, *J. Comput. Chem. 25*, 1605.

37. Farber, G. K., and Petsko, G. A. (1990) The evolution of alpha/beta barrel enzymes, *Trends Biochem. Sci. 15*, 228.

38. Sun, J., and Sampson, N. S. (1999) Understanding protein lids: Kinetic analysis of active hinge mutants in triosephosphate isomerase, *Biochemistry 38*, 11474.

39. Pankiewicz, K. W., Patterson, S. E., Black, P. L., Jayaram, H. N., Risal, D., Goldstein, B. M., et al. (2004) Cofactor mimics as selective inhibitors of NAD-dependent inosine monophosphate dehydrogenase (IMPDH) — The major therapeutic target, *Curr. Med. Chem. 11*, 887.

40. Chong, C. R., Qian, D. Z., Pan, F., Wei, Y., Pili, R., Sullivan, D. J., Jr., and Liu, J. O. (2006) Identification of type 1 inosine monophosphate dehydrogenase as an antiangiogenic drug target, *J. Med. Chem. 49*, 2677.

41. Yamada, Y., Natsumeda, Y., and Weber, G. (1988) Action of the active metabolites of tiazofurin and ribavirin on purified IMP dehydrogenase, *Biochemistry 27*, 2193.

42. Di Bisceglie, A. M., McHutchison, J., and Rice, C. M. (2002) New therapeutic strategies for hepatitis C, *Hepatology 35*, 224.

43. Shad, J. A., and McHutchison, J. G. (2001) Current and future therapies of hepatitis C, *Clin. Liver Dis. 5*, 335.

44. Ikegami, T., Natsumeda, Y., and Weber, G. (1989) Recovery of the activities of IMP dehydrogenase and GMP synthase after treatment with tiazofurin and acivicin in hepatoma cells in vitro, *Adv. Exp. Med. Biol. 253B*, 299.

45. Grifantini, M. (2000) Tiazofurine ICN pharmaceuticals, *Curr. Opin. Investig. Drugs 1*, 257.

46. Franchetti, P., Marchetti, S., Cappellacci, L., Yalowitz, J. A., Jayaram, H. N., Goldstein, B. M., and Grifantini, M. (2001) A new C-nucleoside analogue of tiazofurin: Synthesis and biological evaluation of 2-beta-D-ribofuranosylimidazole-4-carboxamide (imidazofurin), *Bioorg. Med. Chem. Lett. 11*, 67.

47. Pankiewicz, K. W., Lesiak-Watanabe, K. B., Watanabe, K. A., Patterson, S. E., Jayaram, H. N., Yalowitz, J. A., et al. (2002) Novel mycophenolic adenine bis(phosphonate) analogues as potential differentiation agents against human leukemia, *J. Med. Chem. 45*, 703.

48. Sokoloski, J. A., Blair, O. C., Carbone, R., and Sartorelli, A. C. (1989) Induction of the differentiation of synchronized HL-60 leukemia cells by tiazofurin, *Exp. Cell. Res. 182*, 234.

49. Luecke, H., Prosise, G. L., and Whitby, F. G. (1997) Tritrichomonas foetus: A strategy for structure-based inhibitor design of a protozoan inosine-5′-monophosphate dehydrogenase, *Exp. Parasitol. 87*, 203.

50. Wang, C. C., Verham, R., Rice, A., and Tzeng, S. (1983) Purine salvage by Tritrichomonas foetus, *Mol. Biochem. Parasitol. 8*, 325.

51. Verham, R., Meek, T. D., Hedstrom, L., and Wang, C. C. (1987) Purification, characterization, and kinetic analysis of inosine 5'-monophosphate dehydrogenase of Tritrichomonas foetus, *Mol. Biochem. Parasitol. 24*, 1.

52. Hedstrom, L., Cheung, K. S., and Wang, C. C. (1990) A novel mechanism of mycophenolic acid resistance in the protozoan parasite Tritrichomonas foetus, *Biochem. Pharmacol. 39*, 151.

53. Hedstrom, L., and Wang, C. C. (1990) Mycophenolic acid and thiazole adenine dinucleotide inhibition of Tritrichomonas foetus inosine 5'-monophosphate dehydrogenase: Implications on enzyme mechanism, *Biochemistry 29*, 849.

54. Hampton, A., Brox, L. W., and Bayer, M. (1969) Analogs of inosine 5'-phosphate with phosphorus-nitrogen and phosphorus-sulfur bonds. Binding and kinetic studies with inosine 5'-phosphate dehydrogenase, *Biochemistry 8*, 2303.

55. Hampton, A. (1977) 6-Chloropurine ribonucleoside 5'-phosphate, *Methods Enzymol. 46*, 299.

56. Miller, R. L., and Adamczyk, D. L. (1976) Inosine 5'-monophosphate dehydrogenase from sarcoma 180 cells — Substrate and inhibitor specificity, *Biochem. Pharmacol. 25*, 883.

57. Skibo, E. B., and Meyer, R. B., Jr. (1981) Inhibition of inosinic acid dehydrogenase by 8-substituted purine nucleotides, *J. Med. Chem. 24*, 1155.

58. Wong, C. G., and Meyer, R. B., Jr. (1984) Inhibitors of inosinic acid dehydrogenase. 2-Substituted inosinic acids, *J. Med. Chem. 27*, 429.

59. Fuke, S., and Konosu, S. (1991) Taste-active components in some foods: A review of Japanese research, *Physiol. Behav. 49*, 863.

60. Hampton, A., and Nomura, A. (1967) Inosine 5'-phosphate dehydrogenase. Site of inhibition by guanosine 5'-phosphate and of inactivation by 6-chloro- and 6-mercaptopurine ribonucleoside 5'-phosphates, *Biochemistry 6*, 679.

61. Anderson, J. H., and Sartorelli, A. C. (1969) Inhibition of inosinic acid dehydrogenase by 6-chloropurine nucleotide, *Biochem. Pharmacol. 18*, 2735.

62. Brox, L. W., and Hampton, A. (1968) Inosine 5'-phosphate dehydrogenase. Kinetic mechanism and evidence for selective reaction of the 6-chloro analog of inosine 5'-phosphate with a cysteine residue at the inosine 5'-phosphate site, *Biochemistry 7*, 2589.

63. Gilbert, H. J., and Drabble, W. T. (1980) Active-site modification of native and mutant forms of inosine 5'-monophosphate dehydrogenase from Escherichia coli K12, *Biochem. J. 191*, 533.

64. Antonino, L. C., Straub, K., and Wu, J. C. (1994) Probing the active site of human IMP dehydrogenase using halogenated purine riboside 5'-monophosphates and covalent modification reagents, *Biochemistry 33*, 1760.

65. Kerr, K. M., and Hedstrom, L. (1997) The roles of conserved carboxylate residues in IMP dehydrogenase and identification of a transition state analog, *Biochemistry 36*, 13365.

66. Wang, W., and Hedstrom, L. (1998) A potent "fat base" nucleotide inhibitor of IMP dehydrogenase, *Biochemistry 37*, 11949.

67. Goldstein, B. M., Bell, J. E., and Marquez, V. E. (1990) Dehydrogenase binding by tiazofurin anabolites, *J. Med. Chem. 33*, 1123.

68. Robins, R. K., Revankar, G. R., McKernan, P. A., Murray, B. K., Kirsi, J. J., and North, J. A. (1985) The importance of IMP dehydrogenase inhibition in the broad spectrum antiviral activity of ribavirin and selenazofurin, *Adv. Enz. Regul. 24*, 29.

69. Kuttan, R., Robins, R. K., and Saunders, P. P. (1982) Inhibition of inosinate dehydrogenase by metabolites of 2-beta-D-ribofuranosyl thiazole-4-carboxamide, *Biochem. Biophys. Res. Commun. 107*, 862.

70. Fleming, M. A., Chambers, S. P., Connelly, P. R., Nimmesgern, E., Fox, T., Bruzzese, F. J., et al. (1996) Inhibition of IMPDH by mycophenolic acid: Dissection of forward and reverse pathways using capillary electrophoresis, *Biochemistry 35*, 6990.

71. Ji, Y., Gu, J., Makhov, A. M., Griffith, J. D., and Mitchell, B. S. (2006) Regulation of the interaction of inosine monophosphate dehydrogenase with mycophenolic acid by GTP, *J. Biol. Chem. 281*, 206.

72. Jayaram, H. N., Ahluwalia, G. S., Dion, R. L., Gebeyehu, G., Marquez, V. E., Kelley, J. A., et al. (1983) Conversion of 2-beta-D-ribofuranosylselenazole-4-carboxamide to an analogue of NAD with potent IMP dehydrogenase-inhibitory properties, *Biochem. Pharmacol. 32*, 2633.

73. Marquez, V. E., Tseng, C. K., Gebeyehu, G., Cooney, D. A., Ahluwalia, G. S., Kelley, J. A., et al. (1986) Thiazole-4-carboxamide adenine dinucleotide (TAD). Analogues stable to phosphodiesterase hydrolysis, *J. Med. Chem. 29*, 1726.

74. Gosio, B. (1896) Ricerche batteriologichee chimiche sulle alterazioni del mais. *Rev. Igenie Sanita Pubbl. Ann. 7*, 825.

75. Chen, J. L., Gerwick, W. H., Schatzman, R., and Laney, M. (1994) Isorawsonol and related IMP dehydrogenase inhibitors from the tropical green alga Avrainvillea rawsonii, *J. Nat. Prod. 57*, 947.

76. Isshiki, K., Takahashi, Y., Iinuma, H., Naganawa, H., Umezawa, Y., Takeuchi, T., et al. (1987) Synthesis of azepinomycin and its beta-D-ribofuranoside, *J. Antibiot. (Tokyo) 40*, 1461.

77. Jain, J., Almquist, S. J., Shlyakhter, D., and Harding, M. W. (2001) VX-497: A novel, selective IMPDH inhibitor and immunosuppressive agent, *J. Pharm. Sci. 90*, 625.

78. Barnes, B. J., Eakin, A. E., Izydore, R. A., and Hall, I. H. (2001) Implications of selective type II IMP dehydrogenase (IMPDH) inhibition by the 6-ethoxycarbonyl-3,3-disubstituted-1,5-diazabicyclo[3.1.0]hexane-2,4-diones on tumor cell death, *Biochem. Pharmacol. 62*, 91.

79. Dhar, T. G., Shen, Z., Guo, J., Liu, C., Watterson, S. H., Gu, H. H., et al. (2002) Discovery of *N*-[2-[2-[[3-methoxy-4-(5-oxazolyl)phenyl]amino]-5-oxazolyl]phenyl]-*N*-methyl-4-morpholineacetamide as a novel and potent inhibitor of inosine monophosphate dehydrogenase with excellent in vivo activity, *J. Med. Chem. 45*, 2127.

80. Sintchak, M. D., and Nimmesgern, E. (2000) The structure of inosine 5'-monophosphate dehydrogenase and the design of novel inhibitors, *Immunopharmacology 47*, 163.

81. Franklin, T. J., Morris, W. P., Jacobs, V. N., Culbert, E. J., Heys, C. A., Ward, W. H., et al. (1999) A novel series of non-nucleoside inhibitors of inosine 5'-monophosphate dehydrogenase with immunosuppressive activity, *Biochem. Pharmacol. 58*, 867.

82. Watterson, S. H., Chen, P., Zhao, Y., Gu, H. H., Dhar, T. G., Xiao, Z., et al. (2007) Acridone-based inhibitors of inosine 5′-monophosphate dehydrogenase: Discovery and SAR leading to the identification of N-(2-(6-(4-ethylpiperazin-1-yl)pyridin-3-yl)propan-2-yl)-2-fluoro-9-oxo-9,10-dihydroacridine-3-carboxamide (BMS-566419), *J. Med. Chem. 50*, 3730.

83. Heyde, E., and Morrison, J. F. (1976) Studies on inosine monophosphate dehydrogenase. Isotope exchange at equilibrium, *Biochim. Biophys. Acta. 429*, 661.

84. Heyde, E., Nagabhushanam, A., Vonarx, M., and Morrison, J. F. (1976) Studies on inosine monophosphate dehydrogenase. Steady state kinetics, *Biochim. Biophys. Acta 429*, 645.

85. Carr, S. F., Papp, E., Wu, J. C., and Natsumeda, Y. (1993) Characterization of human type I and type II IMP dehydrogenases, *J. Biol. Chem. 268*, 27286.

86. Digits, J. A., and Hedstrom, L. (1999) Species-specific inhibition of inosine 5′-monophosphate dehydrogenase by mycophenolic acid, *Biochemistry 38*, 15388.

87. Wang, W., and Hedstrom, L. (1997) Kinetic mechanism of human inosine 5′-monophosphate dehydrogenase type II: Random addition of substrates and ordered release of products, *Biochemistry 36*, 8479.

88. Xiang, B., and Markham, G. D. (1996) The conformation of inosine 5′-monophosphate (IMP) bound to IMP dehydrogenase determined by transferred nuclear Overhauser effect spectroscopy, *J. Biol. Chem. 271*, 27531.

89. Xiang, B., and Markham, G. D. (1997) Probing the mechanism of inosine monophosphate dehydrogenase with kinetic isotope effects and NMR determination of the hydride transfer stereospecificity, *Arch. Biochem. Biophys. 348*, 378.

90. Cleland, W. W. (1979) Statistical analysis of enzyme kinetic data, *Methods Enzymol. 63*, 103.

91. Cleland, W. W. (1990) Steady state kinetics, in *The Enzymes*, Vol. 19, Academic, New York, p. 99.

92. Antonino, L. C., and Wu, J. C. (1994) Human IMP dehydrogenase catalyzes the dehalogenation of 2-fluoro- and 2-chloroinosine 5′-monophosphate in the absence of NAD, *Biochemistry 33*, 1753.

93. Segel, I. H. (1975) *Enzyme Kinetics. Behavior and Analysis of Rapid Equilibrium and Steady State Enzyme Systems*, Wiley-Interscience, New York.

94. Albery, W. J., and Knowles, J. R. (1976) Evolution of enzyme function and the development of catalytic efficiency, *Biochemistry 15*, 5631.

95. Albery, W. J., and Knowles, J. R. (1977) Efficiency and evolution of enzyme catalysis, *Angew Chem. Int. Ed. Engl. 16*, 285.

95a. Guex, N., and Peitsch, M. C. (1997) SWISS-MODEL and the Swiss-PdbViewer: An environment for comparative protein modeling, *Electrophoresis 18*, 2714.

96. Kerr, K. M., Digits, J. A., Kuperwasser, N., and Hedstrom, L. (2000) Asp338 controls hydride transfer in Escherichia coli IMP dehydrogenase, *Biochemistry 39*, 9804.

97. Link, J. O., and Straub, K. (1996) Trapping of an IMP dehydrogenase-substrate covalent intermediate by mycophenolic acid, *J. Am. Chem. Soc. 118*, 2091.

98. Evans, F. E., and Sarma, R. H. (1974) The tautomeric form of inosine in aqueous solution, *J. Mol. Biol. 89*, 249.

99. Roy, K. B., and Miles, H. T. (1983) Tautomerization and ionization of xanthosine, *Nucleosides Nucleotides 2*, 231.

100. Lin, Y., and Nageswara Rao, B. D. (2000) Structural characterization of adenine nucleotides bound to Escherichia coli adenylate kinase. 2. 31P and 13C relaxation measurements in the presence of cobalt(II) and manganese(II), *Biochemistry 39*, 3647.

101. Futer, O., Sintchak, M. D., Caron, P. R., Nimmesgern, E., DeCenzo, M. T., Livingston, D. J., and Raybuck, S. A. (2002) A mutational analysis of the active site of human type II inosine 5'-monophosphate dehydrogenase, *Biochim. Biophys. Acta 1594*, 27.

102. Schalk-Hihi, C., Zhang, Y. Z., and Markham, G. D. (1998) The conformation of NADH bound to inosine 5'-monophosphate dehydrogenase determined by transferred nuclear Overhauser effect spectroscopy, *Biochemistry 37*, 7608.

103. Arnold, L. J., Jr., You, K., Allison, W. S., and Kaplan, N. O. (1976) Determination of the hydride transfer stereospecificity of nicotinamide adenine dinucleotide linked oxidoreductases by proton magnetic resonance, *Biochemistry 15*, 4844.

104. Cooney, D., Hamel, E., Cohen, M., Kang, G. J., Dalal, M., and Marquez, V. (1987) A simple method for the rapid determination of the stereospecificity of NAD-dependent dehydrogenases applied to mammalian IMP dehydrogenase and bacterial NADH peroxidase, *Biochim. Biophys. Acta 916*, 89.

105. Oppenheimer, N. J., and Handlon, A. L. (1992) Mechanism of NAD–dependent enzymes, in *The Enzymes*, Vol. 20, Academic, New York, p. 454.

106. de Kok, P. M., Beijer, N. A., Buck, H. M., Sluyterman, L. A., and Meijer, E. M. (1988) Molecular mechanics calculation of geometries of NAD+ derivatives, modified in the nicotinamide group, in a ternary complex with horse liver alcohol dehydrogenase, *Eur. J. Biochem. 175*, 581.

107. Li, H., Hallows, W. H., Punzi, J. S., Marquez, V. E., Carrell, H. L., Pankiewicz, K. W., et al. (1994) Crystallographic studies of two alcohol dehydrogenase-bound analogues of thiazole-4-carboxamide adenine dinucleotide (TAD), the active anabolite of the antitumor agent tiazofurin, *Biochemistry 33*, 23.

108. Peters, G. H., Frimurer, T. M., and Olsen, O. H. (1998) Electrostatic evaluation of the signature motif (H/V)CX5R(S/T) in protein-tyrosine phosphatases, *Biochemistry 37*, 5383.

109. Hammes, G. G. (2002) Multiple conformational changes in enzyme catalysis, *Biochemistry 41*, 8221.

110. Joseph, D., Petsko, G. A., and Karplus, M. (1990) Anatomy of a conformational change: Hinged "lid" motion of the triosephosphate isomerase loop, *Science 249*, 1425.

111. Juszczak, L. J., Zhang, Z. Y., Wu, L., Gottfried, D. S., and Eads, D. D. (1997) Rapid loop dynamics of Yersinia protein tyrosine phosphatases, *Biochemistry 36*, 2227.

112. Derreumaux, P., and Schlick, T. (1998) The loop opening/closing motion of the enzyme triosephosphate isomerase, *Biophys. J. 74*, 72.

113. Wang, G. P., Cahill, S. M., Liu, X., Girvin, M. E., and Grubmeyer, C. (1999) Motional dynamics of the catalytic loop in OMP synthase, *Biochemistry 38*, 284.

114. Osborne, M. J., Schnell, J., Benkovic, S. J., Dyson, H. J., and Wright, P. E. (2001) Backbone dynamics in dihydrofolate reductase complexes: Role of loop flexibility in the catalytic mechanism, *Biochemistry 40*, 9846.

115. Day, P., Sharff, A., Parra, L., Cleasby, A., Williams, M., Horer, S., et al. (2007) Structure of a CBS-domain pair from the regulatory gamma1 subunit of human AMPK in complex with AMP and ZMP, *Acta Crystallogr. Sect. D: Biol. Crystallogr. 63*, 587.

116. Falzone, C. J., Wright, P. E., and Benkovic, S. J. (1994) Dynamics of a flexible loop in dihydrofolate reductase from *Escherichia coli* and its implication for catalysis, *Biochemistry 33*, 439.

117. Kempner, E. S. (1993) Movable lobes and flexible loops in proteins. Structural deformations that control biochemical activity, *FEBS Lett. 326*, 4.

118. Pompliano, D. L., Peyman, A., and Knowles, J. R. (1990) Stabilization of a reaction intermediate as a catalytic device: Definition of the functional role of the flexible loop in triosephosphate isomerase, *Biochemistry 29*, 3186.

119. Gerstein, M., Lesk, A. M., and Chothia, C. (1994) Structural mechanisms for domain movements in proteins, *Biochemistry 33*, 6739.

120. Taylor, J. C., and Markham, G. D. (2003) Conformational dynamics of the active site loop of *S*-adenosylmethionine synthetase illuminated by site-directed spin labeling, *Arch. Biochem. Biophys. 415*, 164.

121. Nimmesgern, E., Fox, T., Fleming, M. A., and Thomson, J. A. (1996) Conformational changes and stabilization of inosine 5′-monophosphate dehydrogenase associated with ligand binding and inhibition by mycophenolic acid, *J. Biol. Chem. 271*, 19421.

122. Bruzzese, F. J., and Connelly, P. R. (1997) Allosteric properties of inosine monophosphate dehydrogenase revealed through the thermodynamics of binding of inosine 5′-monophosphate and mycophenolic acid. Temperature dependent heat capacity of binding as a signature of ligand-coupled conformational equilibria, *Biochemistry 36*, 10428.

123. Murphy, K. P., and Freier, E. (1993) Thermodynamics of structural stability and cooperative folding behavior in proteins, *Adv. Protein Chem. 43*, 313.

124. Fisher, H. F. (1988) Unifying model of the thermodynamics of formation of dehydrogenase-ligand complexes, *Adv. Enzymol. 61*, 1.

125. Guillen Schlippe, Y. V., Riera, T. V., Seyedsayamdost, M. R., and Hedstrom, L. (2004) Substitution of the conserved Arg-Tyr dyad selectively disrupts the hydrolysis phase of the IMP dehydrogenase reaction, *Biochemistry 43*, 4511.

126. Guillen Schlippe, Y. V., and Hedstrom, L. (2005) Is Arg418 the catalytic base required for the hydrolysis step of the IMP dehydrogenase reaction? *Biochemistry 44*, 11700.

127. Guillen Schlippe, Y. V., and Hedstrom, L. (2005) Guanidine derivatives rescue the Arg418Ala mutation of Tritrichomonas foetus IMP dehydrogenase, *Biochemistry 44*, 16695.

128. Farazi, T., Leichman, J., Harris, T., Cahoon, M., and Hedstrom, L. (1997) Isolation and characterization of mycophenolic acid-resistant mutants of inosine-5′-monophosphate dehydrogenase, *J. Biol. Chem. 272*, 961.

129. Digits, J. A., and Hedstrom, L. (2000) Drug selectivity is determined by coupling across the NAD+ site of IMP dehydrogenase, *Biochemistry 39*, 1771.

130. Kohler, G. A., Gong, X., Bentink, S., Theiss, S., Pagani, G. M., Agabian, N., and Hedstrom, L. (2005) The functional basis of mycophenolic acid resistance in Candida albicans IMP dehydrogenase, *J. Biol. Chem. 280*, 11295.

131. Hyle, J. W., Shaw, R. J., and Reines, D. (2003) Functional distinctions between IMP dehydrogenase genes in providing mycophenolate resistance and guanine prototrophy to yeast, *J. Biol. Chem. 278*, 28470.

132. Shaw, R. J., Wilson, J. L., Smith, K. T., and Reines, D. (2001) Regulation of an IMP dehydrogenase gene and its overexpression in drug-sensitive transcription elongation mutants of yeast, *J. Biol. Chem. 276*, 32905.

133. McPhillips, C. C., Hyle, J. W., and Reines, D. (2004) Detection of the mycophenolate-inhibited form of IMP dehydrogenase in vivo, *Proc. Natl. Acad. Sci. USA 101*, 12171.

134. Jenks, M. H., and Reines, D. (2005) Dissection of the molecular basis of mycophenolate resistance in Saccharomyces cerevisiae, *Yeast 22*, 1181.

135. Markham, G. D. (2002) Monovalent cation activation of IMP dehydrogenase, in *Inosine Monophosphate Dehydrogenase: A Major Theraputic Target*, Pankiewicz, K. W., and Goldstein, B. M., Eds., Oxford University Press, New York, pp. 169–183.

136. Wu, T. W., and Scrimgeour, K. G. (1973) Properties of inosinic acid dehydrogenase from Bacillus subtilis. II. Kinetic properties, *Can. J. Biochem. 51*, 1391.

137. Beck, J. T., Zhao, S., and Wang, C. C. (1994) Cloning, sequencing, and structural analysis of the DNA encoding inosine monophosphate dehydrogenase (EC 1.1.1.205) from Tritrichomonas foetus, *Exp. Parasitol. 78*, 101.

138. Hupe, D. J., Azzolina, B. A., and Behrens, N. D. (1986) IMP dehydrogenase from the intracellular parasitic protozoan Eimeria tenella and its inhibition by mycophenolic acid, *J. Biol. Chem. 261*, 8363.

139. Glusker, J. P. (1991) Structural aspects of metal liganding to functional groups in proteins, *Adv. Protein Chem. 42*, 1.

140. Larsen, T. M., Laughlin, L. T., Holden, H. M., Rayment, I., and Reed, G. H. (1994) Structure of rabbit muscle pyruvate kinase complexed with Mn^{2+}, K^+, and pyruvate, *Biochemistry 33*, 6301.

141. Toney, M. D., Hohenester, E., Keller, J. W., and Jansonius, J. N. (1995) Structural and mechanistic analysis of two refined crystal structures of the pyridoxal phosphate-dependent enzyme dialkylglycine decarboxylase, *J. Mol. Biol. 245*, 151.

142. Takusagawa, F., Kamitori, S., Misaki, S., and Markham, G. D. (1996) Crystal structure of S-adenosylmethionine synthetase, *J. Biol. Chem. 271*, 136.

143. Rhee, S., Parris, K. D., Ahmed, S. A., Miles, E. W., and Davies, D. R. (1996) Exchange of K+ or Cs+ for Na+ induces local and long-range changes in the three-dimensional structure of the tryptophan synthase alpha2beta2 complex, *Biochemistry 35*, 4211.

144. Antson, A. A., Demidkina, T. V., Gollnick, P., Dauter, Z., von Tersch, R. L., Long, J., et al. (1993) Three-dimensional structure of tyrosine phenol-lyase, *Biochemistry 32*, 4195.

145. Bateman, A. (1997) The structure of a domain common to archaebacteria and the homocystinuria disease protein, *Trends Biochem. Sci. 22*, 12.

146. Ignoul, S., and Eggermont, J. (2005) CBS domains: Structure, function, and pathology in human proteins, *Am. J. Physiol. Cell. Physiol. 289*, C1369.

147. Miller, M. D., Schwarzenbacher, R., von Delft, F., Abdubek, P., Ambing, E., Biorac, T., et al. (2004) Crystal structure of a tandem cystathionine-beta-synthase (CBS) domain protein (TM0935) from Thermotoga maritima at 1.87 A resolution, *Proteins 57*, 213.

148. Townley, R., and Shapiro, L. (2007) Crystal structures of the adenylate sensor from fission yeast AMP-activated protein kinase, *Science 315*, 1726.

149. Meyer, S., Savaresi, S., Forster, I. C., and Dutzler, R. (2007) Nucleotide recognition by the cytoplasmic domain of the human chloride transporter ClC-5, *Nat. Struct. Mol. Biol. 14*, 60.

150. Markovic, S., and Dutzler, R. (2007) The structure of the cytoplasmic domain of the chloride channel ClC-Ka reveals a conserved interaction interface, *Structure 15*, 715.

151. Nimmesgern, E., Black, J., Futer, O., Fulghum, J. R., Chambers, S. P., Brummel, C. L., et al. (1999) Biochemical analysis of the modular enzyme inosine 5'-monophosphate dehydrogenase, *Protein Expr. Purif. 17*, 282.

152. Kennan, A., Aherne, A., Palfi, A., Humphries, M., McKee, A., Stitt, A., et al. (2002) Identification of an IMPDH1 mutation in autosomal dominant retinitis pigmentosa (RP10) revealed following comparative microarray analysis of transcripts derived from retinas of wild-type and Rho($-/-$) mice, *Hum. Mol. Genet. 11*, 547.

153. Bowne, S. J., Sullivan, L. S., Blanton, S. H., Cepko, C. L., Blackshaw, S., Birch, D. G., et al. (2002) Mutations in the inosine monophosphate dehydrogenase 1 gene (IMPDH1) cause the RP10 form of autosomal dominant retinitis pigmentosa, *Hum. Mol. Genet. 11*, 559.

154. Kennan, A., Aherne, A., and Humphries, P. (2005) Light in retinitis pigmentosa, *Trends Genet. 21*, 103.

155. Aherne, A., Kennan, A., Kenna, P. F., McNally, N., Lloyd, D. G., Alberts, I. L., et al. (2004) On the molecular pathology of neurodegeneration in IMPDH1-based retinitis pigmentosa, *Hum. Mol. Genet. 13*, 641.

156. Mortimer, S. E., and Hedstrom, L. (2005) Autosomal dominant retinitis pigmentosa mutations in inosine 5'-monophosphate dehydrogenase type I disrupt nucleic acid binding, *Biochem. J. 390*, 41.

157. Bowne, S. J., Sullivan, L. S., Mortimer, S. E., Hedstrom, L., Zhu, J., Spellicy, C. J., et al. (2006) Spectrum and frequency of mutations in IMPDH1 associated with autosomal dominant retinitis pigmentosa and leber congenital amaurosis, *Invest. Ophthalmol. Vis. Sci. 47*, 34.

158. Grover, S., Fishman, G. A., and Stone, E. M. (2004) A novel IMPDH1 mutation (Arg231Pro) in a family with a severe form of autosomal dominant retinitis pigmentosa, *Ophthalmology 111*, 1910.

159. McLean, J. E., Hamaguchi, N., Belenky, P., Mortimer, S. E., Stanton, M., and Hedstrom, L. (2004) Inosine 5'-monophosphate dehydrogenase binds nucleic acids in vitro and in vivo, *Biochem. J. 379*, 243.

160. Scott, J. W., Hawley, S. A., Green, K. A., Anis, M., Stewart, G., Scullion, G. A., et al. (2004) CBS domains form energy-sensing modules whose binding of adenosine ligands is disrupted by disease mutations, *J. Clin. Invest. 113*, 274.

161. Whitehead, J. P., Simpson, F., Hill, M. M., Thomas, E. C., Connolly, L. M., Collart, F., et al. (2004) Insulin and oleate promote translocation of inosine-5' monophosphate dehydrogenase to lipid bodies, *Traffic 5*, 739.

162. Hardie, D. G. (2003) Minireview: The AMP-activated protein kinase cascade: The key sensor of cellular energy status, *Endocrinology 144*, 5179.

163. Kahn, B. B., Alquier, T., Carling, D., and Hardie, D. G. (2005) AMP-activated protein kinase: Ancient energy gauge provides clues to modern understanding of metabolism, *Cell. Metab. 1*, 15.

164. Jhee, K. H., and Kruger, W. D. (2005) The role of cystathionine beta-synthase in homocysteine metabolism, *Antioxid. Redox Signal. 7*, 813.

165. Biemans-Oldehinkel, E., Mahmood, N. A., and Poolman, B. (2006) A sensor for intracellular ionic strength, *Proc. Natl. Acad. Sci. USA 103*, 10624.

166. Rudolph, M. J., Amodeo, G. A., Iram, S. H., Hong, S. P., Pirino, G., Carlson, M., and Tong, L. (2007) Structure of the Bateman2 domain of yeast Snf4: Dimeric association and relevance for AMP binding, *Structure 15*, 65.

NATURAL PRODUCT GLYCOSYLTRANSFERASES: PROPERTIES AND APPLICATIONS

By GAVIN J. WILLIAMS and JON S. THORSON, *Laboratory for Biosynthetic Chemistry, Pharmaceutical Sciences Division, School of Pharmacy, National Cooperative Drug Discovery Program, University of Wisconsin-Madison, Madison, Wisconsin 53705*

CONTENTS

Advances in Enzymology and Related Areas of Molecular Biology, Volume 76
Edited by Eric J. Toone   Copyright © 2009 by John Wiley & Sons, Inc.

I. Introduction

Natural products continue to serve as the platform in the development of novel therapeutics (1–4). Many therapeutically relevant natural products are decorated with sugars (Figure 1) by an essential class of enzyme known as glycosyltransferases (GTs), the topic of this review. Sugars attached to such natural products by GTs are typically indispensable for the compound's biological activity (5–7) and influence pharmacology and pharmacokinetics, invoke biological specificity, contribute to molecular recognition, and/or even define the precise mechanism of action (1). For example, the disaccharide moiety of the well-known glycopeptide antibiotic vancomycin is critical for activity (8), removal of this disaccharide results in a drastic reduction in antibacterial activity (9), while disaccharide modification retargets the antibiotic to a distinct cellular enzyme and thereby thwarts antibiotic resistance (10). Likewise, appending the naturally occurring tubulin destabilizer colchicine, which normally is not glycosylated, with sugars presents analogs that bind to a distinct tubulin site and remarkably *stabilize* tubulin polymerization (11). In the same vein, the replacement of the natural trisaccharide of the plant natural product digitoxin with variant monosaccharides provides derivatives with markedly improved anticancer activities and greatly reduced cardiotoxicity (12).

The growing appreciation of the importance of natural product sugar moieties has spurred the development of methods for natural product glycosylation and glycodiversification—ranging from new synthetic methodologies to enzyme-intensive approaches. From a synthetic perspective, the sheer complexity of natural products (e.g., calicheamicin γ_1^I, Figure 1, 4) often renders classical chemical glycosylation strategies, and their inherently complex protection/sugar activation/deprotection requirements, impractical for glycodiversification (13–15). However, emerging chemical glycosylation strategies for small molecules such as

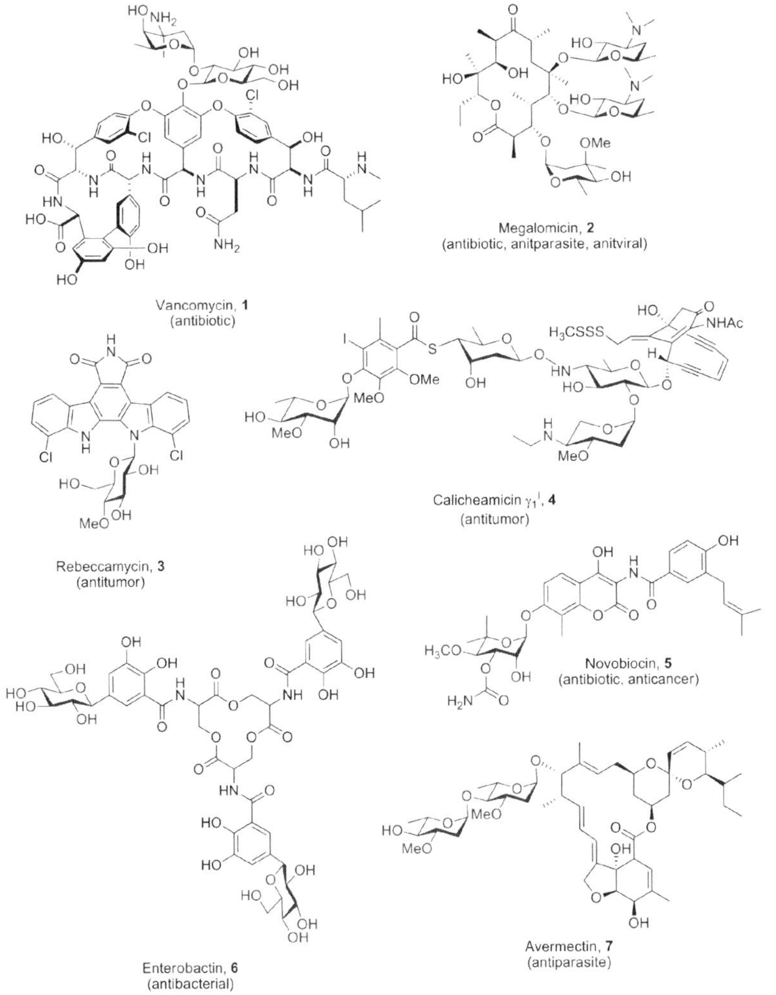

Figure 1.   Representative glycosylated natural products and their biological activities. These examples include O-glycosides (**1**, **2**, **4**, **5**, and **7**), N-glycosides (**3** and **4**), and a C-glycoside (**6**). The glycosyl moieties are highlighted blue. (See insert for color representation.)

O'Doherty's de novo glycosylation strategy (16–18) and the neoglycorandomization strategy developed by Thorson and co-workers (6, 19) are rapidly expanding the application of sugars in small-molecule therapeutic development.

Enzyme-dependent approaches have been accomplished both in vivo (e.g., pathway engineering) and in vitro (e.g., chemoenzymatic glycorandomization). Included under the broad descriptor of *pathway engineering* are gene and gene insertion approaches (Figure 2B) (7, 20, 21), both of which often suffer from technical hurdles associated with producing strain genetic manipulation and a significant reduction in production levels of engineered or shunt metabolites. Some of these limitations have been overcome by the recent development of "plug-and-play" gene cassettes for sugar nucleotide biosynthesis (22–25) which provide for the in vivo synthesis of novel sugar donors to probe glycosylated natural product pathways (26–28). In a similar fashion, in vitro chemoenzymatic glycorandomization also relies upon the inherent promiscuity of natural product GTs and presents a "one-pot" strategy for regio- and stereoselective glycosylation without the need for complex protection/deprotection strategies. However, chemoenzymatic glycorandomization distinguishes itself via the use of engineered kinases and nucleotidyltransferases for the synthesis of diverse sugar nucleotide libraries from a potentially unlimited pool of monosaccharides (Figure 2). To date, chemoenzymatic glycorandomization has been successfully applied toward antibiotic scaffolds (novobiocin, erythromycin, vancomycin) (29–33), anti-cancer models (rebeccamycin/staurosporine/AT2433 and calicheamicin) (33, 34), and antihelmenthics (avermectin/ivermectin) (35).

This review attempts to highlight all reported natural product GTs characterized in vitro to date, with an emphasis on enzymes particularly suited for chemoenzymatic glycorandomization applications. Accordingly, this review first summarizes fundamental GT properties and classification (structural and mechanistic) followed by an emphasis upon the in vitro GT studies to date (as organized by natural product structural class) and, finally, an overview of recent reported efforts to engineer GT specificity, mechanism, and/or catalysis.

II. Glycosyltransferase Sequence, Structure, and Mechanism

A. SEQUENCE CLASSIFICATION

Glycosyltransferases constitute an extremely large and diverse family of enzymes in terms of sequence (36), likely a result of the vast repertoire of donor and acceptor substrates utilized by these enzymes, which in turn is reflected by the structural diversity of glycosides produced (Figure 1).

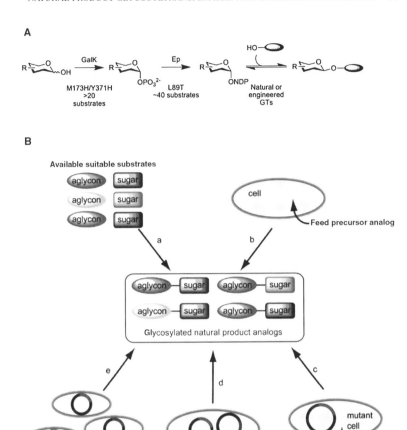

Figure 2. Strategies for glycodiversification of natural products. (A) Scheme for chemoenzymatic glycorandomization. Natural product scaffolds are glycodiversified by the combined action of anomeric kinases (GalK and mutants), nucleotidyltransferases (Ep and mutants), and promiscuous GTs. (B) Several in vivo approaches to natural product glycodiversification. (a) Enzymatic/chemoenzymatic synthesis (e.g., glycorandomization) makes use of isolated sugar biosynthetic enzymes and GTs in conjunction with purified aglycons and NDP donors (or donor precursors). (b) Precursor-directed biosynthesis in a producer organism. (c) Mutasynthesis combines precursor-directed biosynthesis with genetic manipulation (e.g., gene insertion/ deletion/heterologous expression). (d) Metabolic pathway engineering uses gene disruption and heterologous expression to redirect sugar biosynthesis toward a new product. (e) Combinatorial biosynthesis involves the combination of engineered sugar/aglycon biosynthesis pathways to produce hybrid natural products. (See insert for color representation.)

Currently, the CAZY database (http://afmb.cnrs-mrs.fr/~cazy/CAZY/ index.html) contains 23,836 sequences predicted or known GT sequences that are divided into 89 families on the basis of amino acid sequence similarity. In addition to sequence classification, GTs may be classified into two groups according to mechanism of action (37, 38). Inverting enzymes catalyze glycosyl transfer with inversion of configuration at C1 of the donor, while retaining enzymes do so with retention of configuration at C1 of the donor (see Section II.C). The majority of inverting GTs are in CAZY families 1, 2, and 51, while many of the retaining enzymes are in families 4 and 5. Most (if not all) of the natural product GTs discussed in this review are inverting members of family 1 and include enzymes which glycosylate macrolides, nonribosomal peptides, aminocoumarins, enediynes, macrolactams, and aromatic polyketides. These enzymes use an incredible range of donors, including D/L-configured sugars and amino-, methoxy-, and deoxysugars. Interestingly, primitive Archae displays GTs mostly from families 2 and 4, likely the ancestral inverting and retaining families from which most GTs have evolved (39).

B. STRUCTURAL CLASSIFICATION

A total of 60 GTs now have published crystal structures, spanning 28 of the 89 GT families, including both inverting and retaining GTs, at least 9 of which are natural product GTs. In spite of the large variation in substrate specificity and sequence, GTs can be classified mainly into two superfamilies on the basis of structural fold, termed GT-A and GT-B (36, 40–42). The GT-A fold consists of two closely associated Rossman-like $\beta/\alpha/\beta$ domains that form a continuous central sheet exemplified by the structures of the retaining GTs LgtC from *Neisseria meningitidis* (43) and of mannosylglycerate synthase (MGS) (44) (Figure 3A). The GT-A N-terminal domain is largely responsible for binding the nucleotide sugar, while the C-terminal domain predominately contributes to acceptor interactions. All members of the GT-A superfamily contain a signature DXD motif for binding essential divalent metals (45–47). While this divalent metal signature does not exclusively rely upon aspartate, this acid-rich motif always resides in a short loop between the two β-sheets which terminate the first Rossman domain where it interacts with the phosphate groups of the nucleotide donor. Inverting and retaining members of the GT-A superfamily exist, but it remains unclear how a common structural fold can lead to such remarkably divergent mechanistic consequences.

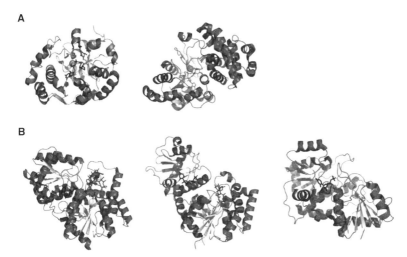

Figure 3.   Structures of representative GTs: (A) GT-A fold; LgtC (*left*) and MGS (*right*); (B) GT-B fold; GtfD (*left*), OleD (*middle*) AviGT4 (*right*). Protein data bank files used were 1GA8, 2B08, 1RRV, 21YF, and 21V3, respectively. In each structure, α-helices are shown in red, β-sheets in yellow, and loops in green. Where applicable, metal ligands are shown cyan, acceptors in dark gray, and donors in blue. (See insert for color representation.)

In the context of the scope of this review, most (if not all) natural product GTs are inverting enzymes and members of the GT-B superfamily. The GT-B fold also consists of two closely associated Rossman-like domains separated by a deep cleft. Within the GT-B structure, the catalytic site is located between the two domains as illustrated by the vancomycin GT GtfD (48), the macrolide GT OleD (49), and the avilamycin GT AviGT4 (39) (Figure 3B). The C-terminal domain, which is largely responsible for nucleotide binding, is highly conserved among many members of the GT-B superfamily. In contrast, this sequence conservation falls off within the N-terminal domains, likely due to the variation in acceptor structures processed by GT-B GTs. Unlike GT-A members, GT-B enzymes are not dependent upon divalent cations, although some GT-B members do show rate enhancements in the presence of divalent metals (50). While the approximate division of donor and acceptor binding between each domain of the GT-B fold [first observed for GtfB (51)] served as the basis for the postulated engineering of hybrid GTs with altered substrate specificity (50), a more recent resolution of the ligand-bound GtfD structure (48) revealed the GT-B C-terminal domain to also contribute essential acceptor interactions. Consistent with this, the successful alteration of GT-B specificity via domain swapping has yet to be reported; however, directed evolution has recently offered advances toward GTs with evolved proficiencies/specificities (see Section IV).

Elucidation of the crystal structures of the diphospholipid donor–dependent peptidoglycan GTs from *Aquifex aelicus* and *Staphylococcus aureus*, involved in the polymerization steps of cell wall biosynthesis, reveals these enzymes to adopt a different fold (almost completely α-helical) from the sugar nucleotide–dependent GT-A and GT-B enzymes (52, 53). These new structures offer a completely unique structural fold to support glycosyltransfer catalysis and may also provide a new target for the discovery of antibiotics.

C. MECHANISMS

The mechanism of the inverting GTs likely involves S_N2 displacement of the nucleotide leaving group, where a general base deprotonates the incoming nucleophile of the acceptor (Figure 4A). Candidates for this key general base in a number of natural product GTs of the GT-B fold have been put forth. For example, mutagenesis of the putative general base (Asp-13) in the vancomycin GT GtfD resulted in a 5500-fold reduction in k_{cat} (48). An identical role for the analogous residue (Asp-13) in GtfA from the chloroeremomycin pathway has been proposed (54). Intriguingly, even though a suitably positioned equivalent to Asp-13 exists within the vancomycin GT GtfB, Asp-332 has been identified as the general base on the basis of mutational analysis (51). Conserved His residues at equivalent positions to the vancomycin GtfD Asp-13 play identical roles in other inverting natural product GT-B enzymes, such as a plant flavonoid GT (55). In the case of the flavonoid GT, it has been proposed that a neighboring serine (Ser-18) and aspartate (Asp-119) take part in a catalytic triad to effect deprotonation of the acceptor hydroxyl (55), similar to the Ser–Asp–His triad of the serine hydrolases. Similar catalytic machinery has been structurally observed in at least one other plant natural product GT (56) and proposed for another on the basis of mutation and analysis of a homology model (57). Cumulatively, no apparent correlation exists between acceptor pK_a values and the participating GT-B inverting general base machinery, and thus the basis for the requirement of a sole aspartate versus a catalytic triad remains unclear (55).

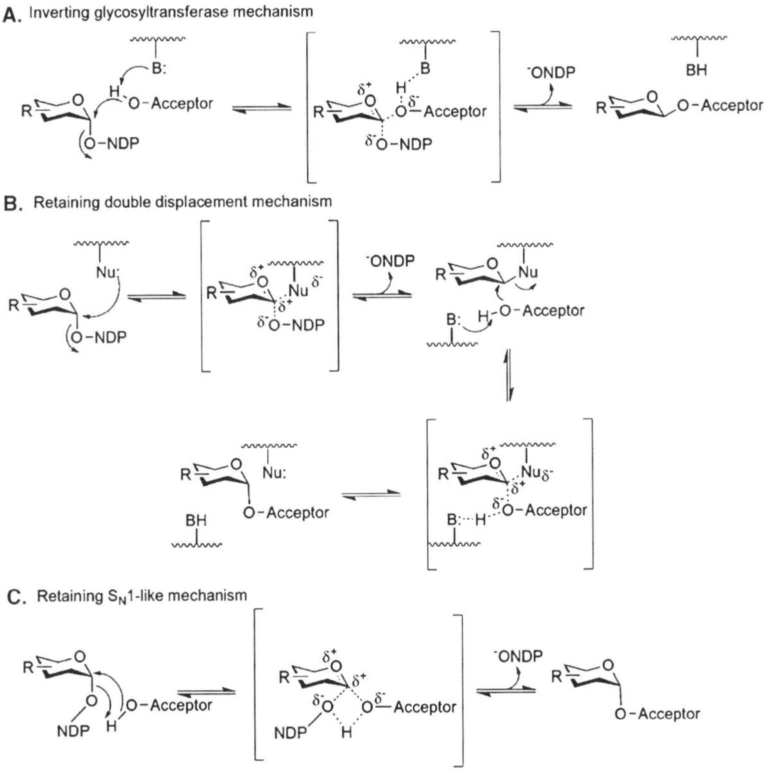

A. Inverting glycosyltransferase mechanism

B. Retaining double displacement mechanism

C. Retaining S_N1-like mechanism

Figure 4.   Mechanisms of GTs.

Inverting enzymes of the GT-A fold also use conserved aspartates and histidines as the general base, albeit in a different sequence location compared to the GT-B enzymes (58, 59). Further, the GT-A inverting enzymes are completely different from their GT-B counterparts, in that the bound divalent metal ion plays the role of a Lewis acid in stabilizing the nucleotide leaving group (47). In addition, the metal ion may also play important roles in determining the sequential mechanism by triggering critical loop movements (60).

In contrast to the inverting GTs, the mechanism of retaining GTs is less well understood, although this is slowly improving with recent structural studies. By analogy to glycosidases, retaining GTs were originally postulated to operate via a double-displacement mechanism requiring two oxocarbenium ion–like transition states (Figure 4B), with the formation of discrete enzyme-bound glycosyl species (61). This mechanism requires the existence of a suitable active-site nucleophile, while a divalent cation plays the role of a Lewis acid and the leaving nucleotide acts as general base to activate the incoming acceptor (62). However, the definitive identification of the catalytic nucleophile for this putative mechanism has remained elusive (43, 63). On this basis, an alternative mechanism involving a single transition state has been proposed (Figure 4C). In the alternative S_Ni-like mechanism, the acceptor itself attacks the leaving group on the same face and has been referred to as a "front-face" mechanism. The arrangement of donor and acceptor in several complexes favors this mechanism (39, 64), and there is at least some chemical precedent in the literature (65). In addition, molecular modeling of the LgtC-retaining GT reaction suggested the front-face mechanism to be more likely based upon quantum mechanical calculations (66). It has been suggested that this mechanism provides a relatively simple route through which inverting GTs have evolved from their retaining counterparts, since only small changes in the position of the acceptor and introduction of a suitable general base would be required to effect this change (44). Perhaps future enzyme engineering efforts will provide experimental support for this proposal.

III. In Vitro Characterization of Natural Product Glycosyltransferases

A. MACROLIDES

Macrolides are macrocyclic lactones, produced as secondary metabolites by the actinomycetes, which typically have one or more deoxy sugars attached (67, 68). Macrolides constitute an important class of antibiotics, as exemplified by erythromycin A (Figure 5, **8**), which was the first macrolide introduced for human clinical use. The classification of the macrolides as first-, second-, and third-generation compounds illustrates the continuing effort and interest in generating new macrolides for therapeutic use. The first-generation compounds were the natural compounds isolated by fermentation and included in addition to **8**, carbomycin (Figure 5, **10**). The second-generation compounds were semisynthetic derivatives of the natural products and included clarithromycin **9** (Biaxin, Abbot) and azithromycin **11** (Zithromax, Pfizer). Finally, the third-generation macrolides, termed the ketolides, were developed due to a sudden emergence of resistance in the

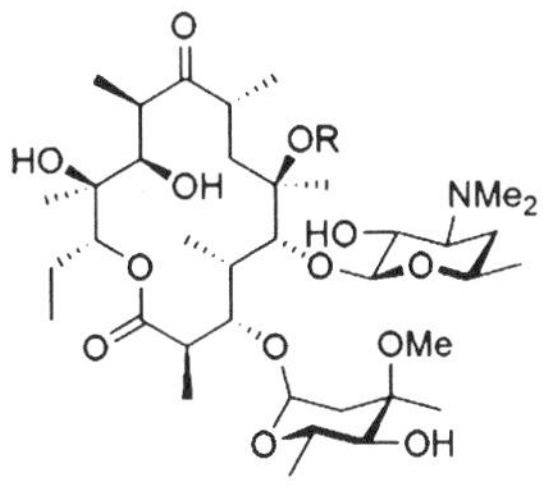

8, R=H
9, R=Me

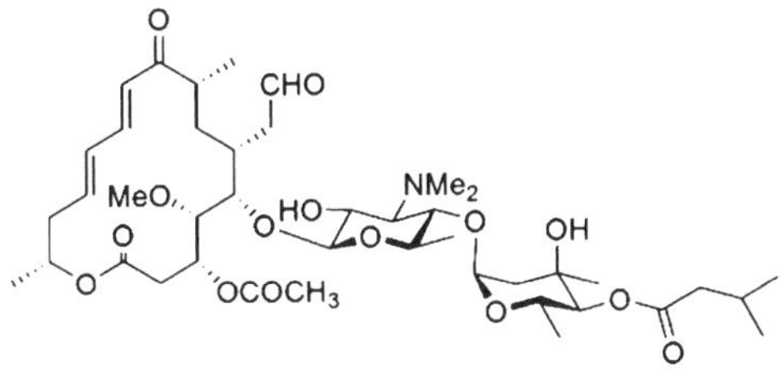

10

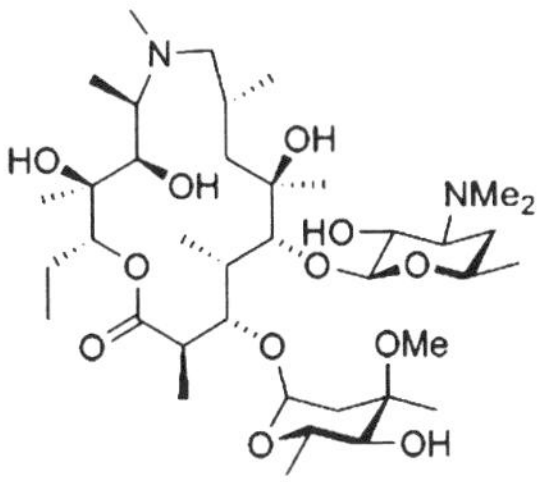

11

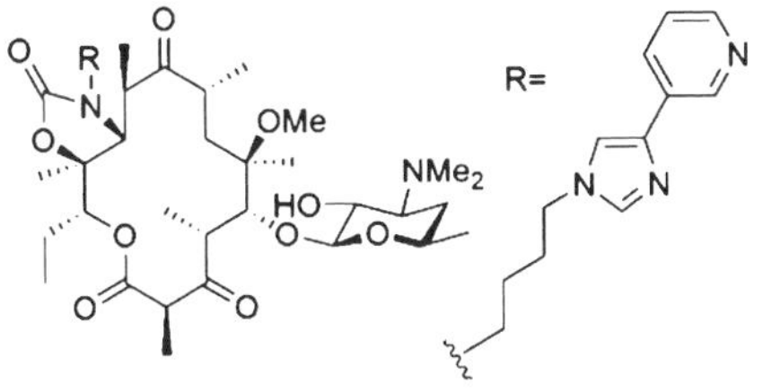

12

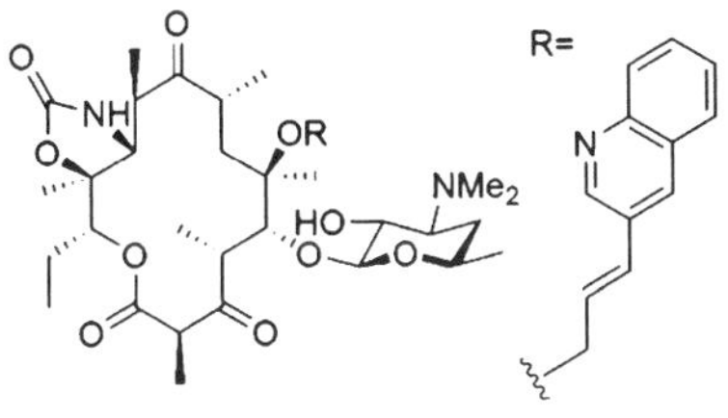

13

Figure 5.   Structures of clinically used macrolides.

1980s and 1990s. The only third-generation macrolide currently in clinical use is telithromycin **12** (Ketek, Avantis), although cethromycin **13** (ABT-773, Abbott) is currently undergoing phase III clinical development. Most macrolides act by targeting bacterial translation and as such are rather broad-spectrum antibiotics. However, the naturally occurring megalomicins produced by *Micromonospora megalomicea* (Figure 1, **2**) are a result of the addition of L-megosamine to the erythromycins. This additional sugar residue remarkably imparts the ability to inhibit protein trafficking in the golgi (69) and, as a result, the megalomicins function as effective antiparasite/antiviral agents (70).

Dedicated macrolide glycosyltransferases are responsible for attaching the deoxy sugars to the macrocyclic lactone aglycones. In most cases, the sugar moiety is critical for bioactivity. In fact, the recent structural elucidation of the 50S ribosomal subunit in complex with various macrolides highlights the critical recognition imparted by the macrolide deoxy sugars (71, 72). Thus, alteration of the macrolide sugars may present a strategy to circumvent antibiotic resistance. As described herein, the recent availability of a number of macrolide GTs provides exciting opportunities to begin altering the glycosyl moiety of these natural products for the discovery of new anti-infective agents.

1. Methymycin: DesVII–DesVIII

In *Streptomyces venezuelae*, a single GT DesVII is responsible for the transfer of desosamine to the acceptors **14** and **15** to produce **16** and **17** (Figure 6). Subsequent hydroxylation by PikC is responsible for producing the 12- and 14-membered macrolides methymycin/neomethymycin (**18/19**) and narbomycin/pikromycin (**17/20**), respectively. The interesting role of a single enzyme glycosylating both 12- and 14-membered macrolides was first proposed on the basis of a single GT gene (*desVII*) found within the pikromycin cluster (73). Intriguingly, the function of an adjacent gene (known to be essential for glycosylation) in the sugar nucleotide subcluster, *desVIII*, could not be assigned, and the activity of purified heterologously expressed DesVII initially could not be reconstituted (74). On this basis, Liu and co-workers investigated whether DesVII required an additional protein partner (such as DesVIII) for in vitro activity. In this landmark study, enzymatic turnover of both the 12- and 14-membered aglycones 10-deoxymethynolide (**14**) and narbonolide (**15**), respectively, with dTDP-desosamine (**21**) was observed when DesVII was incubated in the presence of partially purified DesVIII at pH 9 (74). This finding demonstrated the first in vitro activity of a macrolide GT and the first example of a GT requiring an additional protein for activity. The authors also noted that DesVIII homologs were present in a number of amino-sugar-containing antibiotic biosynthetic gene clusters and proposed the auxiliary protein may be a general requirement for natural product amino-sugar GTs.

Extending these studies, Borisova et al. examined the aglycon and sugar substrate specificity of the DesVII–DesVIII pair (75). This study, which

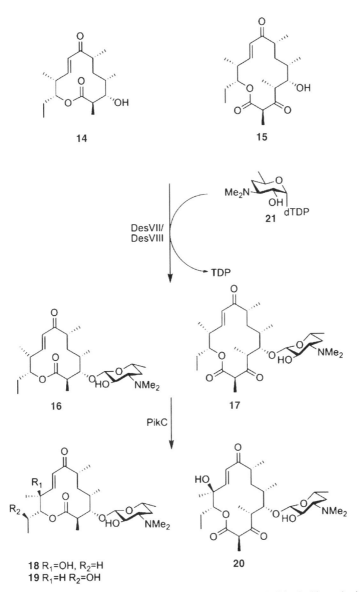

Figure 6.   Combined action of DesVII–DesVIII and hydroxylase PikC in the biosynthesis of methymycin (**18**), neomethymycin (**19**), narbomycin (**17**), and pikromycin (**20**).

employed 24 sugar nucleotides and 5 potential acceptors, stands as one of the broadest interrogations of macrolide GT specificity to date. A total of 7 (Figure 7, **21** and **25–30**) DesVII–DesVIII sugar nucleotide substrates were identified, many of which were unique to those previously discovered via in vivo studies. Although both D- and L-configured sugars were accepted, DesVII/DesVIII displayed a stringent requirement for 6-deoxysugars. Remarkably, sugar nucleotides **25–30** led to products with the natural acceptors **14** and **15**, the hydroxylated variants of 10-deoxymethynolide (**22** and **23**), and the 16-membered macrolide tylactone (**24**). In total, 19

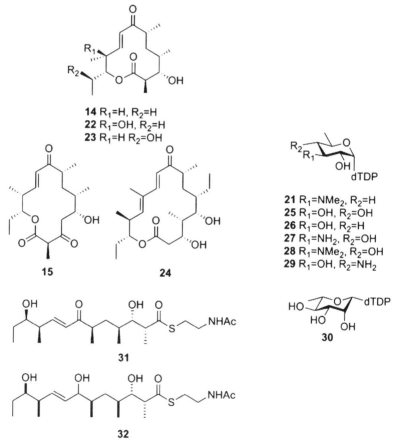

14 R$_1$=H, R$_2$=H
22 R$_1$=OH, R$_2$=H
23 R$_1$=H R$_2$=OH

21 R$_1$=NMe$_2$, R$_2$=H
25 R$_1$=OH, R$_2$=OH
26 R$_1$=OH, R$_2$=H
27 R$_1$=NH$_2$, R$_2$=OH
28 R$_1$=NMe$_2$, R$_2$=OH
29 R$_1$=OH, R$_2$=NH$_2$

15   24

30

31

32

Figure 7.   Acceptors and dTDP donor substrates of DesVII–DesVIII.

previously unreported glycosylated macrolides were produced by the promiscuity of the DesVII–DesVIII pair. In combination with earlier in vivo approaches (76, 77), these results clearly indicate that DesVIII possesses an incredibly broad substrate tolerance. Moreover, in notable contrast to the previously postulated amino-sugar carrier role for the auxiliary protein (74), this study revealed DesVIII to be essential for activity with any sugar nucleotide–aglycon combination.

More recently, DesVII–DesVIII has been shown to regioselectively glycosylate linear analogs of the cyclic macrolide aglycon (Figure 7, **31** and **32**) using **21** as donor, albeit with relatively low efficiency. While this study suggests the substitution patterns near the glycosylation site to be the major DesVII recognition elements, rather than the ring structure (78), the biological significance of this result is yet to be determined.

2.   Erythromycin: EryBV and EryCIII–EryCII

The erythromycin polyketide core 6-deoxyerythronolide B (Figure 8, **33**) undergoes several tailoring steps to produce the mature erythromycins **8** and

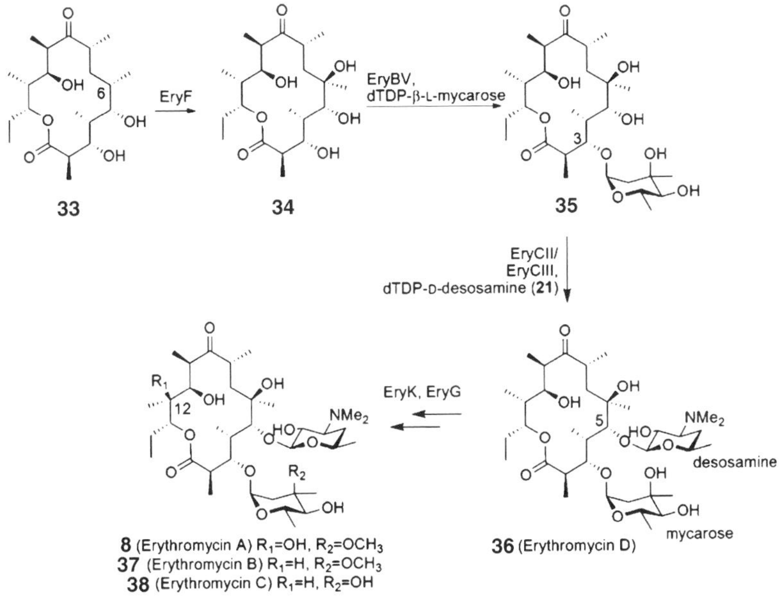

Figure 8.   Action of the GTs EryBV and EryCIII in the biosynthesis of erythromycins A–D (**8, 36–38**).

36–38. These include hydroxylation at C6 and C12 (EryF/EryK respectively), *O*-methylation (EryG), and glycosylation at C3 (mycarose) and C5 (desosamine). The first glycosylation entails the transfer of dTDP-β-L-mycarose to the aglycon **34** (Figure 8). Early gene disruption studies revealed *eryBV* as encoding the candidate GT for this reaction (79). While the availability of the putative EryBV substrate, dTDP-β-L-mycarose, hampered the initial EryBV in vitro studies, the reversibility of the EryBV reaction allowed for the first demonstration of in vitro activity (30). Specifically, EryBV catalyzed the excision of mycarose from readily available 3α-mycarosylerythronolide B (**35**) to give the deglycosylated aglycone **34** and dTDP-β-L-mycarose **39** in the presence of excess dTDP (Figure 9A). Using the same strategy, erythromycins A–D (Figure 8, **8** and **36–38**) were also assayed with EryBV for the dTDP-mediated reverse catalysis. Both erythromycins B (**37**) and D (**36**) led to the EryBV-catalyzed production of free aglycon and dTDP-β-L-mycarose.

The use of this reverse reaction was extended further by transferring the in situ generated TDP-β-L-mycarose (**39**) from 3α-mycarosylerythronolide B (**35**) to an alternative acceptor, 6-deoxyerythronolide B (**34**) (Figure 9A). This *aglycon exchange* reaction produced the new macrolide **39** in 22% yield in a one-pot reaction. While EryBV displayed stringent sugar nucleotide specificity, dTDP-β-L-oleandrose (Figure 9B, **41**), harvested from avermectin **7** by a reverse AveBI catalyzed reaction (see Section III.A.4), was successfully transferred to the erythronolides **33** and **34** by EryBV to provide oleandrosides **42** and **43** (as well as monoglycoside **44**) in good yield. Notably, in contrast to other macrolide GTs (DesVII/VIII and EryCIII/CII), EryBV displayed no requirement for an auxiliary protein for activity.

Early in vivo studies revealed the second glycosylation in erythromycin biosynthesis to be encoded by *eryCIII* (80, 81). Shortly after the initial DesVII–DesVIII in vitro study (described previously in this section), Walker and co-workers established an in vitro system for the EryCIII-catalyzed production of erythromycin D (**36**) from mycarosyl erythronolide B (αMEB) (Figure 10, **35**) and dTDP-D-desosamine (**21**) (82). Soluble EryCIII for this initial study could only be obtained by coexpression with the GroEL/ES chaperone complex, and the purified enzyme rapidly lost activity. In a follow-up investigation, EryCIII in vitro activity was found to be greatly enhanced in the presence of the DesVIII homolog from the erythromycin pathway (EryCII) (83). For this work, coexpression of native *eryCII* in conjunction with N-His$_6$-EryCIII allowed for the affinity purification of EryCIII devoid of any detectable EryCII. Remarkably, EryCIII prepared in this manner was

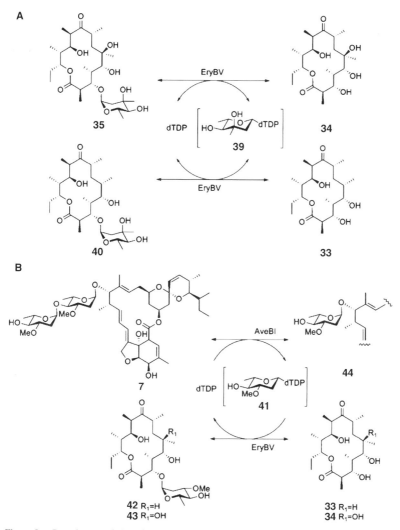

Figure 9.   Reactions carried out by EryBV in vitro. (A) Reverse reaction catalyzed by EryBV and in situ transfer of excised **39** to an alternative acceptor **33**. (B) AveBI-catalyzed generation of **41** and EryBV transfer to acceptors **33** and **34**.

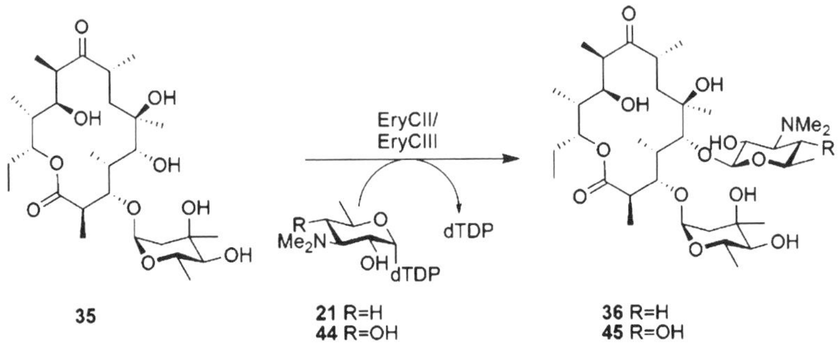

Figure 10.   Reactions catalyzed by the EryCIII–EryCII pair.

active and stable, implicating an "activating" role for EryCII and related proteins (such as DesVIII). To further illustrate this point, the EryCII homolog AknT from the aklavinone biosynthetic pathway was also found to activate EryCII in vitro. To date, EryCIII–EryCII has been interrogated with only two sugar nucleotide donors (dTDP-D-desosamine, **21**; dTDP-D-mycaminose, **44**), both of which were accepted as substrates. Whether alternative sugar donors (such as the 6-deoxy dTDP-sugars described above for DesVIII–DesVII) will be accepted by the EryCIII–EryCII pair remains to be determined. The precise nature of the GT "activation" by DesVIII homologs also remains a mystery.

3.   Oleandomycin: OleI and OleD

The gene cluster encoding for the biosynthesis of oleandomycin (Figure 11, **46**), a macrolide antibiotic produced by *Streptomyces antibioticus*, contains three GT genes (*oleI, oleG1, oleG2*) (84, 85). OleI and the functionally related OleD [encoded by a gene outside the *S. antibioticus ole* gene cluster (86)], catalyze the transfer of glucose from UDP-Glc (**47**) to the desosamine 2′-OH of **46** giving glucosylated oleandomycin **48** (85, 87). This oleandomycin glucosylation prohibits the macrolide from binding to host ribosomes and a secreted glucosidase (OleR) reactivates the macrolide as it exits the producing organism (85). Substrate specificity is the distinguishing feature between OleI and OleD—OleI displays very stringent specificity toward oleandomycin (85) while OleD can glucosylate an array of macrolides including erythromycin (**8**), carbomycin (**10**), and tylosin (Figure 11, **49**). Macrolide glycosylating activities similar to OleD (often referred to as MGT) have been found in crude extracts of at least 15 different *Streptomyces* species (88).

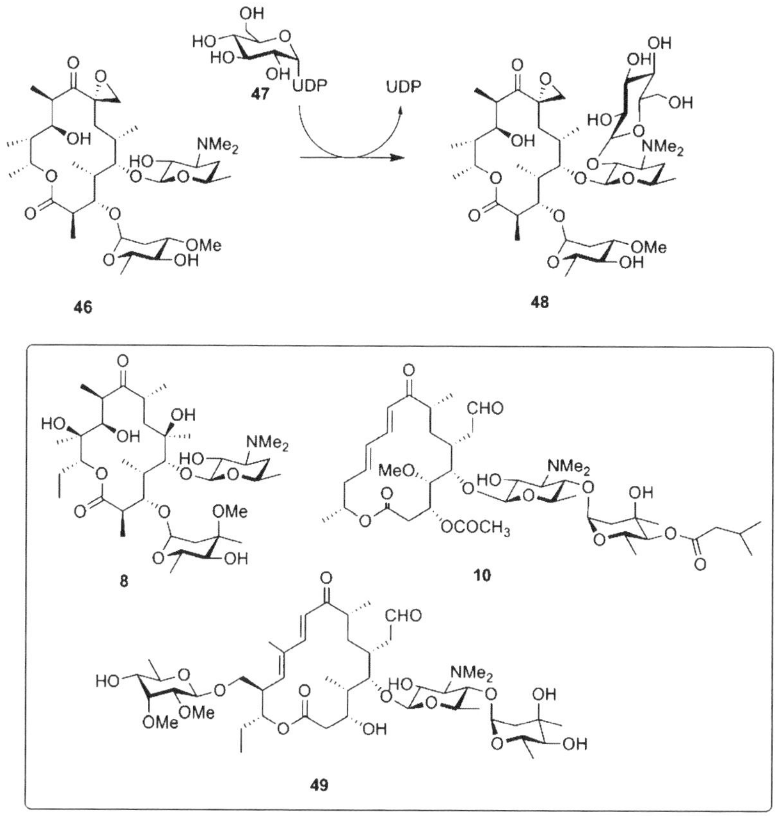

Figure 11. Glucosylation of oleandomycin **46** by OleD. Alternative macrolide substrates are shown boxed.

A detailed kinetic analysis of OleI revealed that the substrates bind in a compulsory order (oleandomycin followed by UDP-glucose) to form a ternary complex in which the glucosyltransfer reaction takes place (87). UDP is the first product to be released, followed by glycosylated oleandomycin. This represents one of the very few kinetic studies of a natural product GT to date. A similarly rigorous kinetic analysis of OleD revealed an ordered mechanism consistent with OleI (89). Substrate specificity studies were recently extended by an examination of the donor and acceptor specificity of OleD, OleI, and MGT using a mass spectrometry–based "green/amber/red" assay (90, 91). A total of 64 acceptors were tested with UDP-glucose (**47**) as

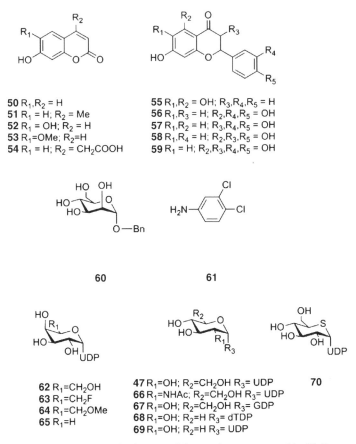

50 R_1,R_2 = H
51 R_1 = H; R_2 = Me
52 R_1 = OH; R_2 = H
53 R_1=OMe; R_2=H
54 R_1 = H; R_2 = CH_2COOH

55 R_1,R_2 = OH; R_3,R_4,R_5 = H
56 R_1,R_3 = H; R_2,R_4,R_5 = OH
57 R_1,R_2 = H; R_3,R_4,R_5 = OH
58 R_1,R_4 = H; R_2,R_3,R_5 = OH
59 R_1 = H; R_2,R_3,R_4,R_5 = OH

60   61

62 R_1=CH_2OH
63 R_1=CH_2F
64 R_1=CH_2OMe
65 R_1=H

47 R_1=OH; R_2=CH_2OH R_3= UDP
66 R_1=NHAc; R_2=CH_2OH R_3= UDP
67 R_1=OH; R_2=CH_2OH R_3= GDP
68 R_1=OH; R_2=H R_3= dTDP
69 R_1=OH; R_2=H R_3= UDP

70

Figure 12.   Unnatural aglycone and donor substrates accepted by OleD.

donor in the presence of each GT. A mass consistent with OleD-catalyzed glucosylation was observed with 15 of these, including flavanols, coumarins, and other small aromatics (Figure 12, **50–61**), one benzyl monosaccharide **60**, 3,4-dichloroaniline (**61**, possibly forming a C–N glycosidic bond), and macrolides **46**, **8**, and **49**. Using the same assay, a total of 18 potential donors were also tested as substrates. Out of this set, a mass consistent with OleD-glycosylation of at least one aglycon was observed with 10 of the sugar nucleotides tested (**62–70**). Notably, the putative OleD donor specificities were highly acceptor dependent. For example, UDP-D-galactose (**62**) (but not

UDP-6-deoxy-6-fluoro-D-galactose, **63**; or UDP-L-arabinose, **65**) led to detectable glycosylation of only kaempferol **58** (of 11 acceptors tested), while UDP-6-deoxy-6-fluoro-D-galactose (**63**) or UDP-L-arabinose (**65**) (but not UDP-D-galactose, **62**) led to detectable glycosylation of the natural macrolide **46**. In a similar manner, a mass consistent with macrolide xylosylation from UDP-D-xylose (**69**) was observed with erythromycin (**8**) but not tylosin (**49**). Consistent with the recent structural elucidation of OleD and OleI (49) this study implicates a complex interaction between donor and acceptor domains which may limit the potential utility of OleD, and possibly GTs in general (92), for true combinatorial natural product diversification. As described in Section IV.C, the recent directed evolution of OleD has led to remarkably improved variants in terms of catalytic proficiency and substrate promiscuity.

4. Avermectin

Avermectins (e.g., Figure 1, **7**) are 16-membered macrocyclic lactones produced by *Streptomyces avermectinius*. These compounds, as well as the related ivermectins, target the γ-aminobutyric acid related chloride ion channels unique to nematodes, insects, ticks, and arachnids with little cytotoxicity (93). Avermectins have been used widely as veterinary antiparasitic agents and more recently in clinical applications including treatment of onchocerciasis and lymphatic filariasis (93). Although avermectins are structurally similar to macrolide antibiotics, avermectins have no antibacterial or antifungal activities (94). Interestingly, the existence of a single GT gene (*aveBI*) within the avermectin gene cluster (95), and corresponding in vivo studies (96, 97), suggested AveBI to catalyze an iterative addition of two oleandrose moieties in a stepwise manner en route to the avermectins. Other iterative natural product GTs include LanGT1 and LanGT4 involved in the biosynthesis of the landomycin oligosaccharide (98). While the availability of the putative AveBI substrate, dTDP-β-L-oleandrose, hampered the initial AveBI in vitro studies, the reversibility of the AveBI reaction allowed for the first demonstration of in vitro activity (35). In a manner similar to that described for EryBV (Section III.A.2), incubation of commercially available avermectin B1$_a$ (**7**) and dTDP with AveBI led to the facile production of dTDP-β-L-oleandrose (**41**) and monoglycosylated variant **44** (Figure 13A). Moreover, in situ transfer of oleandrose from **41** to ivermectin **72** provided mono- and diglycosylated **73** and **74**, respectively (Figure 13B) and demonstrated unequivocally that AveBI catalyzed a stepwise tandem glycosylation

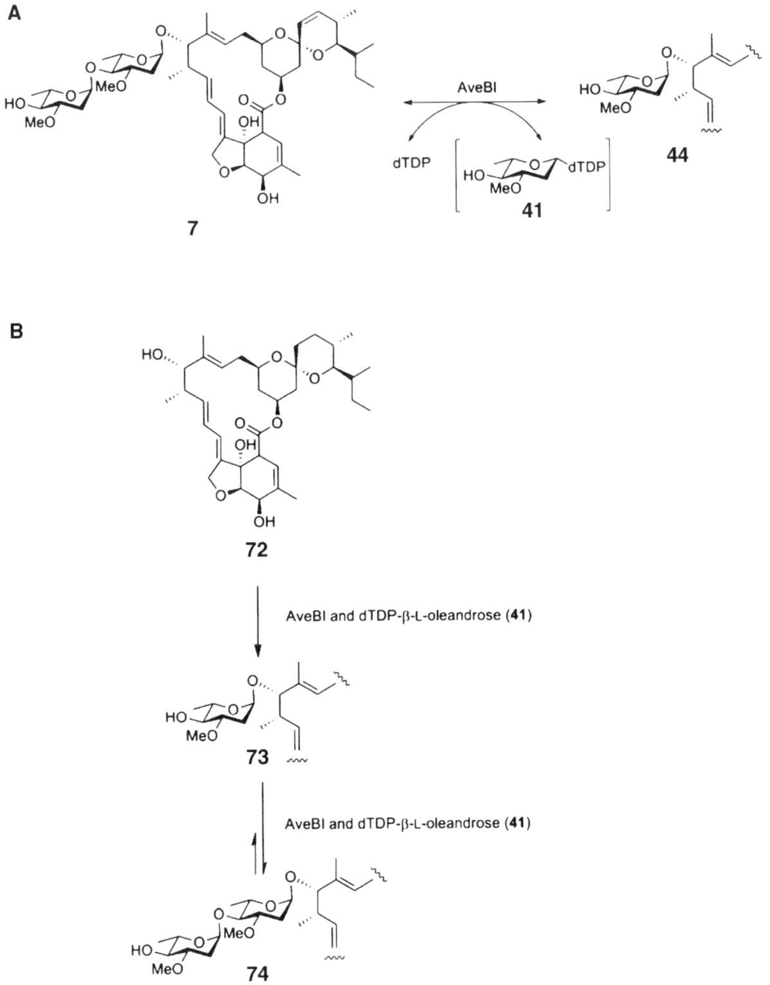

Figure 13.   Reactions catalyzed by AveBI. (A) Reverse reaction from avermectin **7** to generate dTDP-sugar **41**. (B) Tandem addition of **41** to ivermectin **72**.

event. The enzyme was found to be moderately promiscuous toward alternative sugar nucleotide donors as 10 variant sugar nucleotides (comprised of both D- and L-sugars) led to the successful glycosylation of both the fully deglycosylated and monoglycosylated avermectin aglycons in varying yields. In agreement with in vivo studies (97), AveBI could not support the *tandem* addition of D-configured sugars. However, AveBI reversibility in conjunction with a diverse sugar nucleotide library led to the production of 50 avermectin variants and clearly showcased the utility of GTs for rapidly diversifying natural products with therapeutic potential.

5. Sorangicin: SorF

Sorangicins are macrolide antibiotics produced by *Sorangium cellulosum* So ce12 and are among the most potent myxobacterial antibiotics identified to date (99). Sorangicins inhibit bacterial RNA-polymerase by binding to the same site as rifampicin (100), a structurally distinct antituberculosis agent. Interestingly, sorangicin can effectively inhibit rifampicin-resistant RNA-polymerase (101) wherein the conformational flexibility of sorangicin, in contrast to the rigidity of rifampicin, has been proposed as the mechanism by which sorangicin bypasses rifampicin resistance (100). Recently, a gene designated *sorF* was identified within the putative *S. cellulosum* sorangicin biosynthetic gene cluster (102). The corresponding enzyme (SorF) was demonstrated to glucosylate sorangicin aglycon in vitro using both UDP-Glc and dTDP-Glc with similar efficiency (Figure 14) (102). In addition, the enzyme also accepted several other sugar nucleotide donors to allow for the

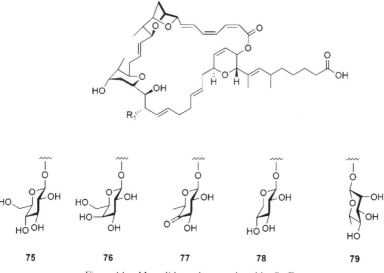

Figure 14.   Macrolide analogs produced by SorF.

production of a total of five sorangicin analogs (Figure 14, **75–79**) and stands among the growing number of natural product GTs capable of accepting both D- and L-sugars. On the basis of this study, the authors speculated SorF may function in a self-resistance capacity similar to macrolide-associated OleD/ OleI (Section III.A.3). SorF displays homology to members of the CAZY family GT-1-inverting GTs and has been proposed to adopt a GT-B fold (102).

B. NONRIBOSOMAL PEPTIDES

Vancomycin and teicoplanin (Figure 15, **1** and **80**) are the only two glycopeptide antibiotics licensed for clinical use in humans. Both are the

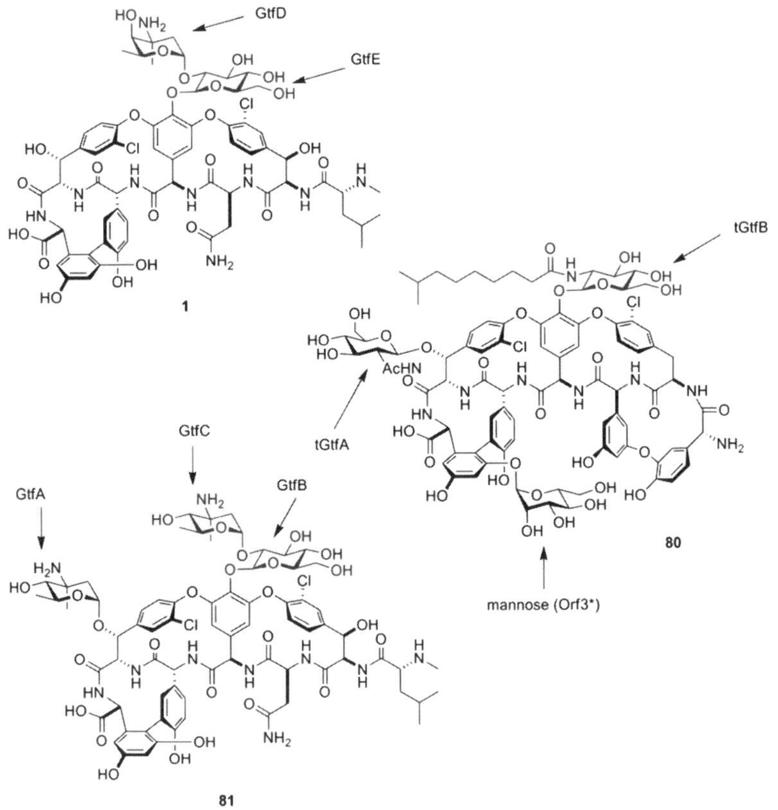

Figure 15.   Naturally occurring glycopeptide antibiotics. Sugar moieties are highlighted blue. (See insert for color representation.)

antibiotics of last resort for treatment of life-threatening infections caused by methicillin-resistant gram-positive bacteria. The rapid global emergence of vancomycin-resistant enterococci (VRE) and vancomycin-resistant *Staphylococcus aureus* (VRSA) (103, 104) has spurred efforts for the development of novel glycopeptides with improved activity resistant strains (8, 105). Vancomycin and related glycopeptides kill bacteria by specifically binding to the *N*-acyl-D-Ala-D-Ala termini of uncrosslinked lipid-PP-disaccharide-pentapeptides (10, 106, 107). This cell wall precursor binding inhibits the transglycosylase/transpeptidase activities required for cell wall rigidity and thus renders bacteria susceptible to lysis by osmotic pressure. The L-vancosaminyl-D-glucose disaccharide of **1** is not directly involved in binding D-Ala-D-Ala, but three lines of evidence implicate the critical role of these carbohydrates in bioactivity. First, removal of the disaccharide provides an aglycon with markedly reduced antibacterial activity (9, 108). Second, N-alkylation of the terminal vancosamine of **1** with a hydrophobic group dramatically increases activity against **1**-resistant strains (109, 110). Finally, some **1** analogs containing synthetically modified carbohydrates operate via a mechanism distinct to **1** (8). Three successful second-generation glycopeptide analogs—oritavancin (Figure 16, **82**), dalbavancin (Figure 16, **83**), and telavancin (Figure 16, **84**)—have capitalized upon these general principles (10, 106, 107, 111). In general, the attachment of sugars bearing straight-chain "lipids" invoke the ability of the glycopeptides to evade induction of resistance (active against VanB) while sugars bearing large aromatics (such as chlorobiphenyl) lead to the inhibition of the bacterial transglycosylase (active against VanA and VanB).

1. Vancomycin and Chloroeremomycin

Preliminary annotation of the biosynthetic clusters for chloroeremycin (Figure 15, **81**) and vancomycin (Figure 15, **1**) formed the basis for in vitro studies of the chloroeremomycin GTs GtfA, GtfB, and GtfC and vancomycin GTs GtfD and GtfE (9, 112). All five of these GTs have now been cloned, over expressed, and characterized in vitro. As expected, GtfB was shown to efficiently glycosylate vancomycin aglycon **85** (Figure 17) to produce vancomycin pseudoaglycon, **86** (113). Kinetic analysis revealed a k_{cat} of $17 \, \text{min}^{-1}$ for transfer of glucose to **85**, and a K_m of $1.3 \, \text{mM}$ for the nucleotide sugar, and activity was severely inhibited at higher concentrations of natural acceptor **85**. The chloroeremomycin homolog GtfE displayed a k_{cat} around threefold higher than GtfB and interestingly no substrate inhibition by

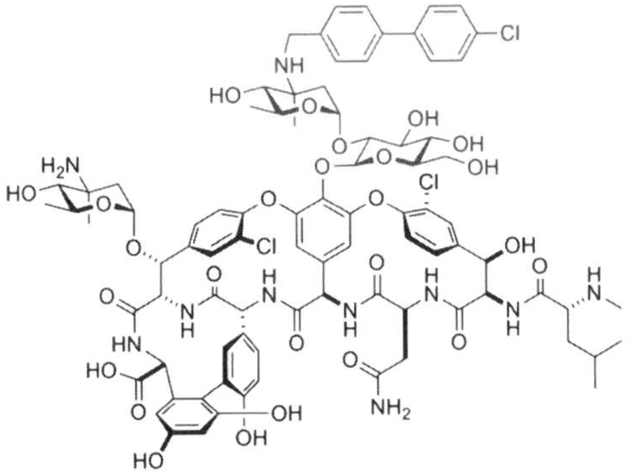

82

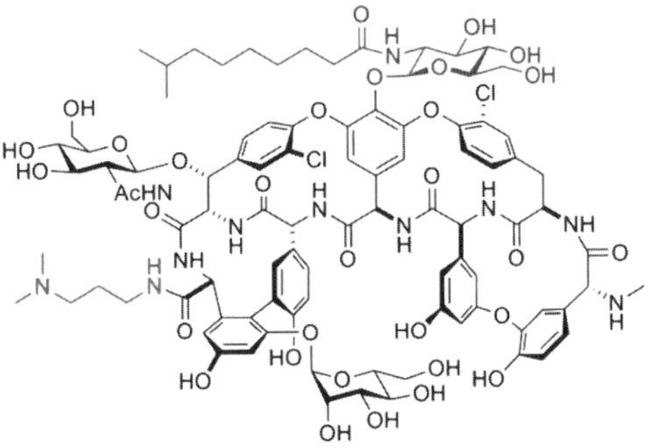

83

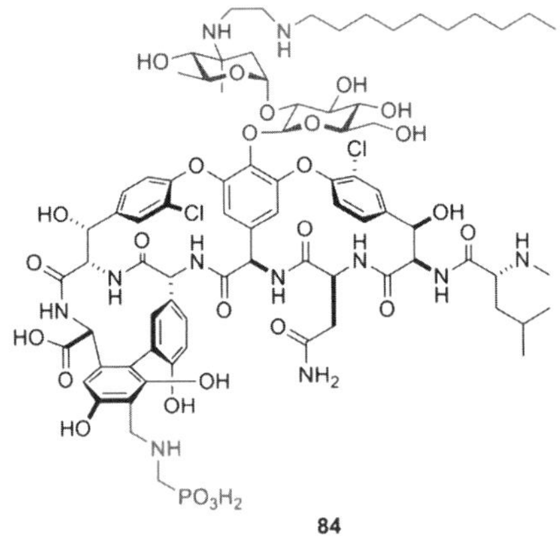

84

Figure 16. Second-generation glycopeptide antibiotics. Sugar moieties are highlighted blue and structural deviations from the parent natural glycopeptide highlighted red. (See insert for color representation.)

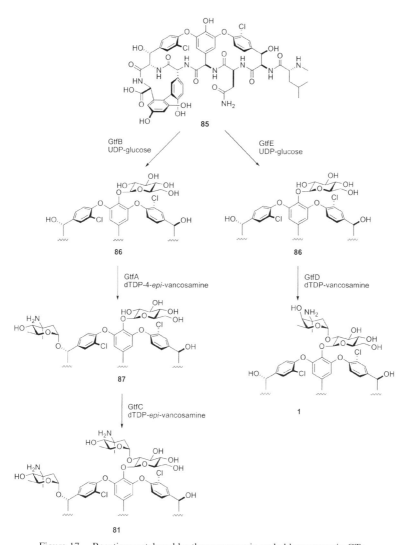

Figure 17. Reactions catalyzed by the vancomycin and chloroeremycin GTs.

aglycon **85**. Subsequently, GtfC was demonstrated to transfer 4-*epi*-vancosamine to vancomycin pseudoaglycon **86** (113) while GtfD could utilize either dTDP-L-vancosamine or UDP-4-*epi*-L-vancosamine in analogous reactions (92, 113). Finally, characterization of GtfA allowed reconstitution of chloroeremomycin from the heptapeptide aglycon **85** (114). Kinetic compari-son of GtfA-catalyzed transfer of 4-*epi*-vancosamine to the heptapeptide aglycon **85**, monoglycosylated vancomycin pseudoaglycon **86**, and disaccha-ride-containing **87** revealed that GtfA preferentially glycosylates the mono-glycosylated aglycon, establishing the glycosylation pathway for the matura-tion of chloroeremomycin to be in the order GtfB, GtfA, and GtfC (Figure 17) (114). More recently, both the GtfD- and GtfE-catalyzed reactions were demonstrated to be reversible in a manner similar to that described for macrolides (Section III.A.2) and avermectins (Section III.A.4) (33).The structures of three glycopeptide GTs, GtfA (54), GtfB (51), and GtfD (48), have been reported and provide a template from which to delineate the roles of active-site residues in substrate binding and catalysis.

Expanding upon the initial substrate specificity work of Solenberg et al. (9), a systematic evaluation of GtfE sugar nucleotide specificity revealed the enzyme to accept 31 out of 33 natural and unnatural NDP-sugars tested— including configurational changes at C2, C3, and C4 and altered functionality at C2, C3, C4, and C6 (31, 108). These included derivatives with reactive handles for downstream chemoselective modification and allowed the ability to rapidly generate 18 analogs from which derivatives with increased activity against *Staphylococcus aureus* and *Enterococcus faecium* were identified (31). Kinetic analysis with a subset of these unnatural NDP-sugars indicated very similar apparent K_m values and a surprisingly narrow range of turnover numbers with GtfE (108). For example, all four regioisomers of NDP-amino-glucose gave K_m values in the range 0.7–1.2 and k_{cat} values in the range 1.3–7.1. Moreover, some unnatural monoglycosylated vancomycins were subse-quently accepted by GtfD with UDP-L-4-*epi*-vancosamine as donor, yielding disaccharide products with alterations in each of the two sugars (108).

In a similar manner, a small set of dTDP-2-deoxy-β-L-sugars were synthesized to probe the promiscuity of GtfA, C, and D (92). Out of 10 donors tested, GtfA, C, and D could only tolerate the removal of the sugar 3-*C*-methyl group while only GtfC and GtfD could also tolerate inversion of the sugar C4 stereochemistry in the presence of **86** as the acceptor. A subsequent detailed kinetic analysis of these reactions demonstrated the turnover numbers for natural donor–acceptor substrate pairs to differ from those of unnatural donor–acceptor substrate pairs by as much as ~40,000-fold. The cumulative kinetic and substrate promiscuity analyses of GtfA–E also led to the hypothesis that GT promiscuity directly correlates with enzyme proficiency (k_{cat}) (92). However, while the proficiency–promiscuity correlation may hold true for this particular series of glycopeptide GTs, examples of highly proficient natural product GTs, with notably stringent specificities are also known. For example, as highlighted in Section III.C.1, NovM displays a k_{cat} of 300 min^{-1} with native substrates (115) (~2-fold higher than the most proficient glycopeptide GT, GtfD) but a remarkably narrow sugar nucleotide specificity (29).

2.   Teicoplanin

Although teicoplanin (Figure 15, **80**) presents subtle variations in both the glycopeptide backbone and sugar substitutions, the latter is predominately attributed to the unique potency, pharmacokinetics, and incapacity to induce glycopeptide resistance (116). The most notable contribution in this regard derives from the unique N-acyl-glucosamine found at PheGly$_4$ (the equivalent disaccharide attachment point in vancomycin). In contrast to vancomycin, teicoplanin also contains N-acetyl-glucosamine at β-OH-Tyr$_6$ and mannose at the C3-hydroxyl of 3,5-dihydroxyphenylglycine$_7$. As anticipated, annotation of the *tei* cluster revealed two gene products (tGtfA and tGtfB) with significant homology to GtfA and GtfB, respectively, in addition to a putative mannosyltransferase (Orf3*) (117). Heterologous expression, purification, and characterization confirmed the roles of tGtfA and tGtfB as UDP-N-acetyl-glucosaminyltransferases (117). Interestingly, tGtfB could also transfer glucosamine and several N-acyl-glucosamine analogs to residue 4 of the teicoplanin aglycon but could not accept the vancomycin aglycon as substrate. Subsequent kinetic characterization of tGtfA and tGtfB established the kinetic order of action of these GTs in the maturation of teicoplanin. In this study, tGtfB was shown to act first by transferring GlcNAc to position 4 of teicoplanin aglycon **88**, to yield 4-GlcNac-AGT **89** (Figure 18). Following deacetylation to provide 4-(2′-aminoglucose)-AGT **90** (118), tGtfA was found to catalyze the addition of GlcNAc to position 6 of 4-(2′-aminoglucose)-AGT **90** to give the bisglycoside **91**. While acyltransferase tAtf likely acts to condense decanoyl-CoA with 4-(2′-aminoglucose)-AGT **90**, giving **92** (postulated as the native in vivo substrate for tGtfA), this has not been established. Subsequent mannosylation by Orf3* affords the mature glycopeptide **80**.

3.   Enterobactins

Enterobactin (Ent) (Figure 19A, **94**) is a prototypical siderophore—low-molecular weight molecules with high affinity for Fe(III) excreted by bacteria. These compounds serve to sequester valuable Fe(III), which

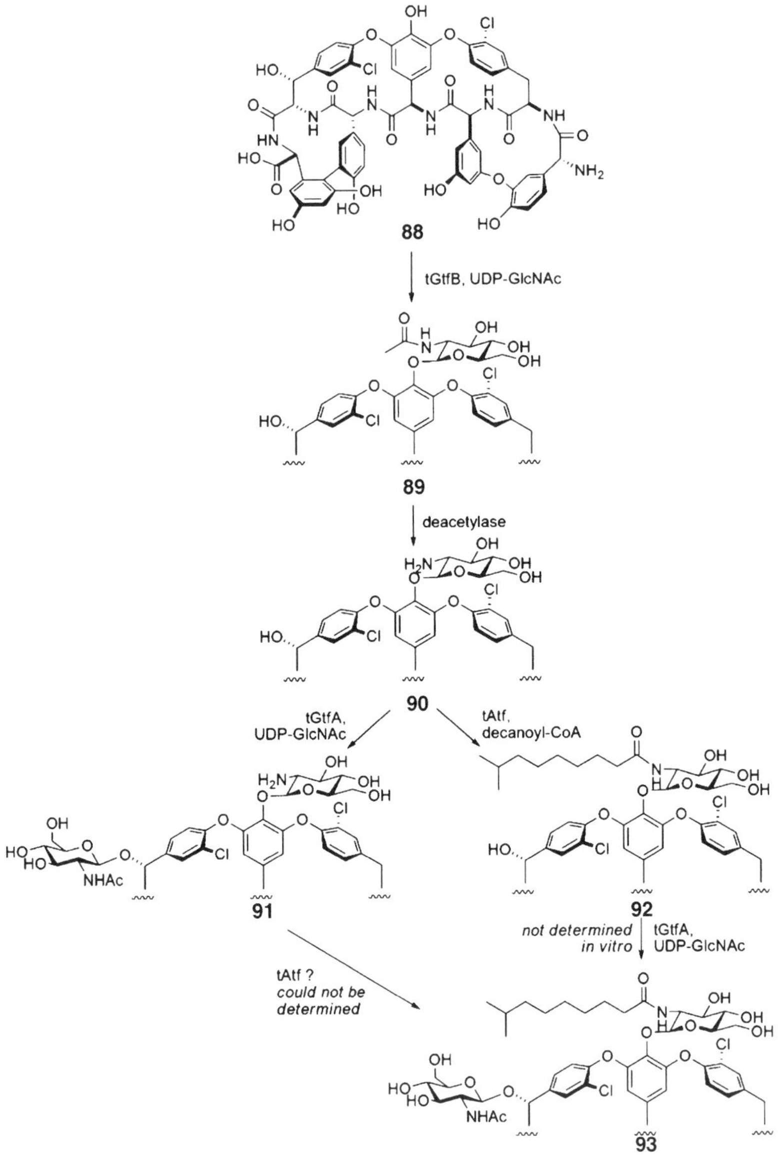

Figure 18.   Proposed biosynthetic pathway for teicoplanin.

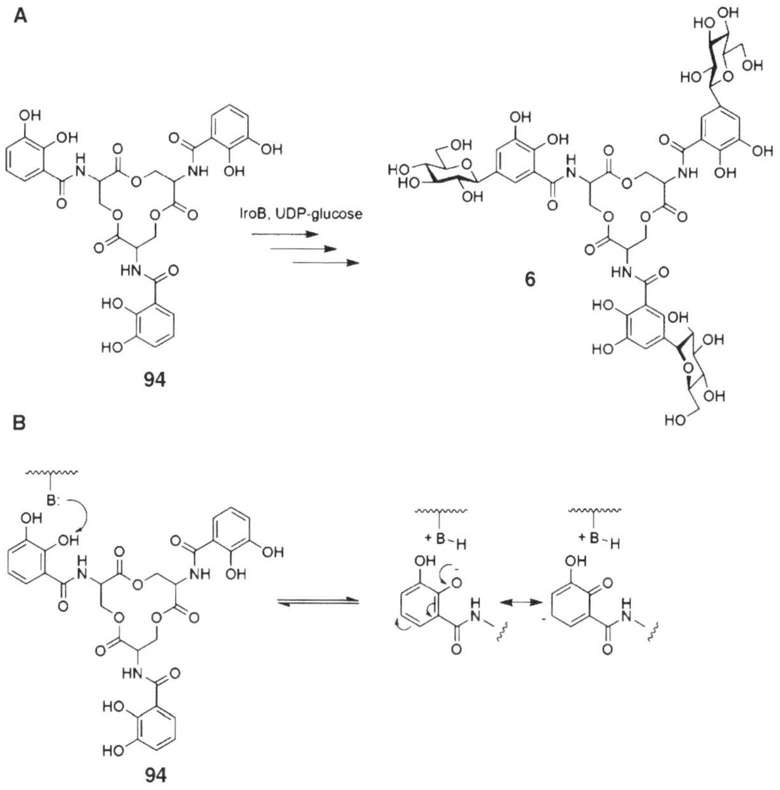

Figure 19.   Glucosylation of Ent **94** by IroB. (A) Conversion of Ent to triglucoside **6**. (B) Postulated mechanism catalyzed by IroB.

otherwise is present in very low concentrations in bacterial pathogens that thrive under highly competitive physiological environments. Enterobactin is comprised of a trimeric macrolactone of N-(2,3-dihydroxybenzoyl) serine, and certain pathogenic strains of *Escherichia coli* and *Salmonella enterica* produce C-glucosides of **94** called salmochelins (encoded by the *iroA* cluster). The putative salmochelin C-glycosyltransferase gene *iroB* was recently overexpressed in *E. coli* and the purified protein product characterized in vitro. This study confirmed IroB to catalyze glucosyl transfer from UDP-glucose to one, two, or three of the 2,3-hydroxybenzoyl units of Ent (**94**) to yield monoglucosyl-C-Ent, diglucosyl-C-Ent, and triglucosyl-C-Ent (**6**).

A subsequent attempt to trap the putative C5 carbanion (Figure 19B) via in vitro deuterium exchange was unable to conclusively delineate among the C-GT mechanisms (direct *C*-glycosylation or *O*-glycosylation followed by O-to-C migration) previously proposed for the hedamycin C-GTs HedJ/HedL (119, 120) and urdamycin UrdGT2 (119, 120). While in vitro characterization of these latter enzymes is completely lacking, in vivo studies support the direct *C*-glycosylation mechanism (120), and notably, the structure of UrdGT2 was also very recently elucidated and found to adopt a GT-B fold (121).

C. AMINOCOUMARINS

Novobiocin (**5**), clorobiocin (**95**), coumermycin A$_1$ (**96**), and simocyclinone D8 (**97**) are aminocoumarin antibiotics produced from various *Streptomyces* strains (Figure 20). These antibiotics inhibit ATP hydrolysis in the bacterial type II DNA topoisomerase DNA gyrase B subunit (122). Although in general the coumarin antibiotics display poor solubility and pharmacological properties, potent activity against methicillin-resistant *S. aureus* bacterial strains has renewed recent interest. Moreover, novobiocin also binds to the *C*-terminal ATP binding site of Hsp90 (123). The known client proteins of Hsp90 (124) suggest that inhibition of the Hsp90 protein-folding machinery could lead to disruption of signaling nodes involved in all six "hallmarks of cancer" (125), making novobiocin and other Hsp90 inhibitors very promising cancer therapeutics. Thus, the possibility of altering the noviose moiety of aminocoumarin antibiotics offers potential to create both improved antibacterial and anticancer leads.

1. Noviobiocin

The aminocoumarin consists of a common core structure of 3-amino-4-hydroxy-coumarin, a prenylated 4-hydroxybenzoic acid moiety, and a deoxy sugar (noviose) essential for biological activity. Interrogation of the *nov* gene cluster in *Streptomyces spheroides* revealed a putative GT gene (*novM*) (126). This gene was subsequently overexpressed in *E. coli* (115) and the corresponding purified *N*-His$_6$-NovM catalyzed the transfer of noviose to the noviobiocic acid aglycon (Figure 21, **98**) using dTDP donor **99** in vitro with a k_{cat} of 300 min^{-1} (one of the fastest GTs examined to date). NovM also displayed reduced in vitro activity with a number of simplified coumarin analogs. For example, cyclonovobiocic acid (Figure 22, **101**) was accepted

5

95

96

97

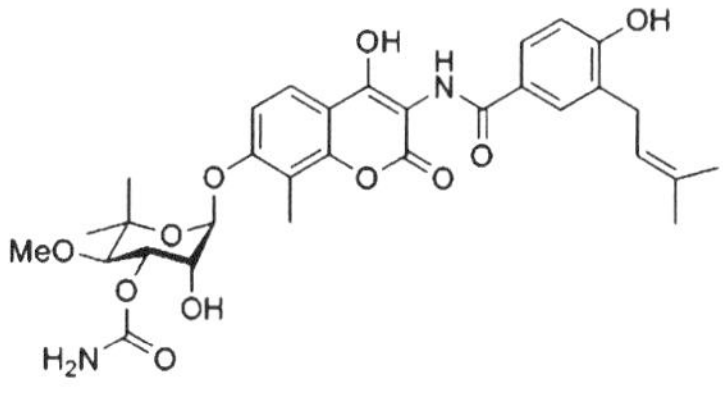

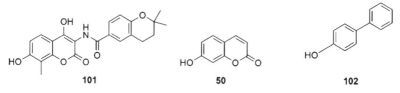

Figure 20. Aminocoumarin antibiotics.

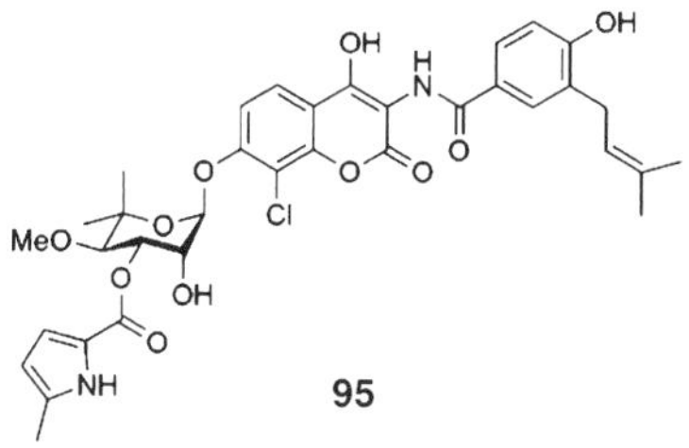

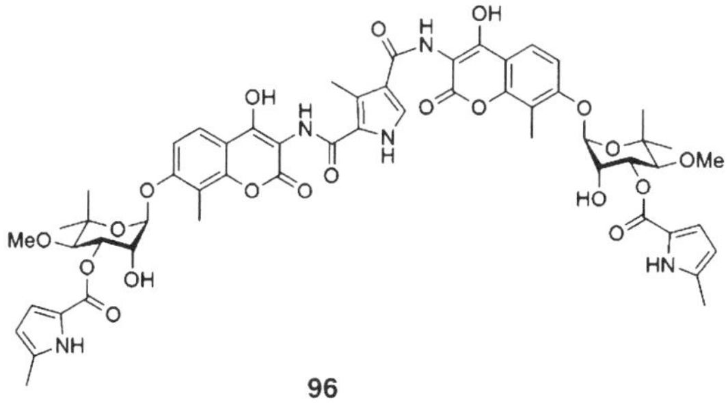

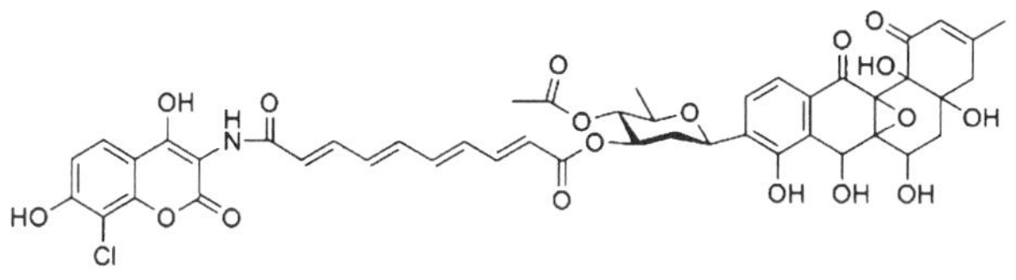

with only a slight (2.5-fold) decrease in k_{cat} while dramatic reductions were observed upon removal of the coumarin 3-amido and 4-hydroxyl (e.g., **50** led to a reduction in k_{cat}/K_m of ~6 orders of magnitude) or using simple phenols (e.g., **102** led to a reduction in k_{cat}/K_m of ~7 orders of magnitude).

101

50

102

Figure 21. Reaction catalyzed by NovM.

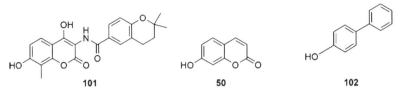

Figure 22.   Alternate acceptors for the NovM-catalyzed transfer of TDP-L-noviose **99**.

Interestingly, dTDP-L-rhamnose, which only differs from the natural donor dTDP-noviose at C5, was an inhibitor of the NovM reaction. A more exhaustive search for potential NovM donors using a library of 34 UDP- and dTDP-sugars revealed a surprisingly narrow donor specificity (29). Specifically, only dTDP-6-deoxy-D-glucose, UDP-6-deoxy-D-glucose, UDP-6-deoxy-xylose, and dTDP-4-keto-6-deoxy-D-glucose were detectable substrates. Remarkably, NovM also supports the transfer of dTDP-noviose to the simocyclinone **97** (Figure 20), an antibiotic that contains an aromatic angucycline polyketide glycoside linked to an aminocoumarin via a tetraene dicarboxylic acid moiety (127).

2. Coumermycin

Coumermycin A_1 (Figure 20, **96**) is a dimeric aminocoumarin antibiotic consisting of a 3-methyl-2,4-dicarboxypyrrole scaffold linking two decorated noviosyl-aminocoumarin moieties, which is the most potent of the aminocoumarin antibiotics. A brief in vitro characterization of purified CouM GT confirmed the ability of this enzyme to add both noviosyl substituents, the order of which remains to be determined (128). The enzyme could also glycosylate a minimalized monoamide.

D. ENEDIYNES

1. Calicheamicin

Enediynes are among nature's most ingenious and potent antibiotics. Structurally, these antibiotics comprise an enediyne core of two acetylenic groups conjugated to a double or incipient double bond within a 9- or 10-membered ring (129). Enediynes exert their cytotoxic activity via DNA damage (130) and, in some cases, chromosomal cleavage specificity has been

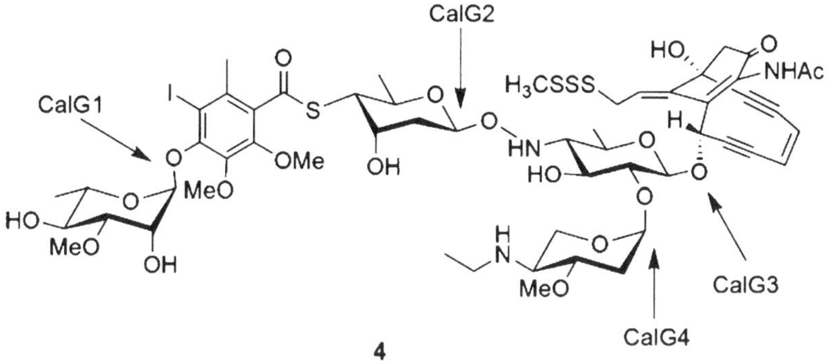

Figure 23. Structure of calicheamicin (**4**) and designation of the GTs responsible for construction of the glycosidic bonds.

observed (131). The incredible potency of enediynes has fueled their development as anticancer agents, and conjugation of a flagship member of the enediyne family (calicheamicin, Figure 23, **4**) to tumor-specific monoclonal antibodies (as exemplified by Mylotarg, the first approved antibody cytotoxin drug) was recently approved for the treatment of acute myelogenous leukemia (132).

Calicheamicin consists of an aryltetrasaccharide (composed of unique deoxy-sugars and an aromatic unit which target the molecule to DNA), the enediyne aglycon core (the active "warhead" which leads to a reactive diradical species), and an allylic trisulfide (the trigger initiated under reducing conditions). The recent identification and annotation of the calicheamicin biosynthetic gene cluster revealed four putative GT genes (*calG1*, *calG2*, *calG3*, and *calG4*) for the attachment of each of the four sugars in calicheamicin (Figure 23) (133). Overproduction of N-His$_{10}$-CalG1 in $E.$ $coli$ and subsequent substrate specificity studies revealed CalG1 to accept 10 of 22 dTDP-sugar donors examined, including dTDP-β-L-rhamnose (Figure 24, **30**) and TDP-α-D-sugars (**25–29, 105–110**) (33) with **111** as acceptor. More importantly, the in vitro studies of CalG1 lead to the ground-breaking discovery that the CalG1 reaction was readily reversible and presented prototypes for "sugar exchange" and "aglycon exchange" reactions (33). As an example of sugar exchange, CalG1 could catalyze the TDP-dependent removal of the 3-O-methylrhamnose moiety of calicheamicin α_3^{I} (**103**) and in the presence of endogenous donor **106** provided a product in which the

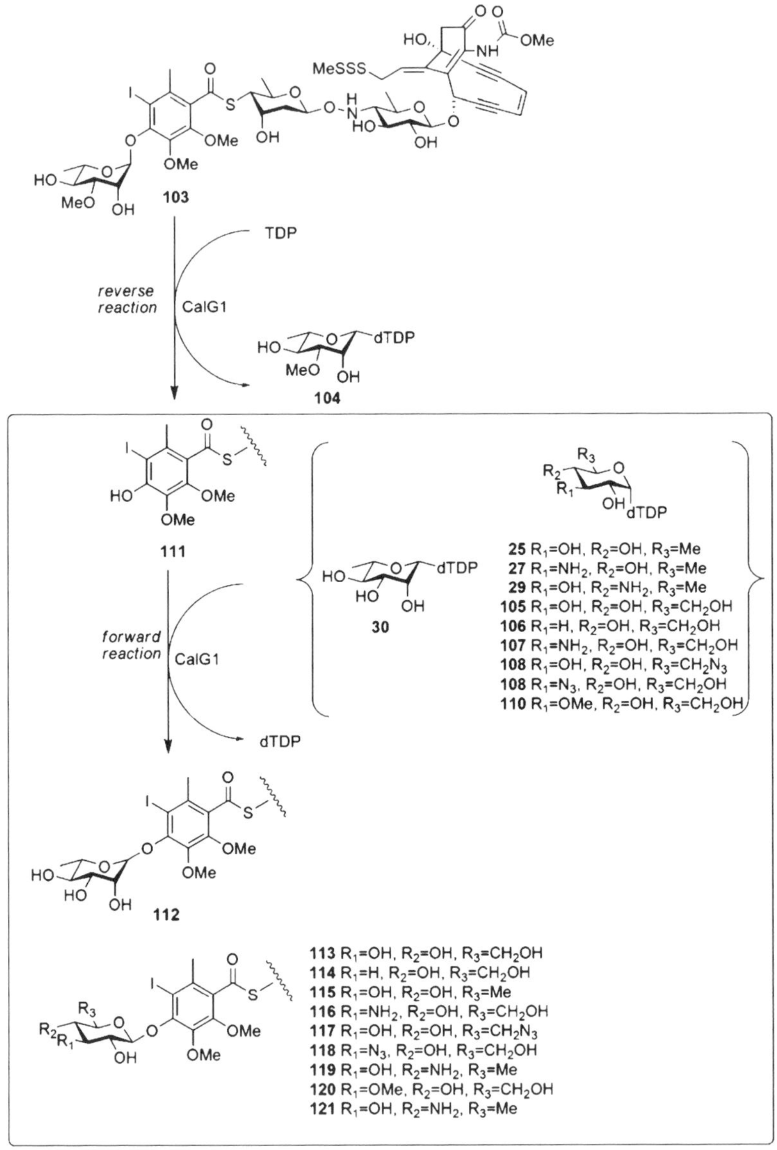

Figure 24.   In vitro sugar exchange reactions catalyzed by CalG1.

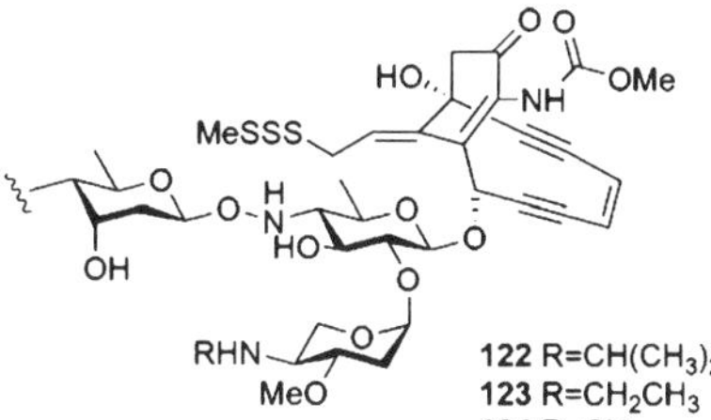

122 R=CH(CH_3)$_2$
123 R=CH_2CH_3
124 R=CH_3

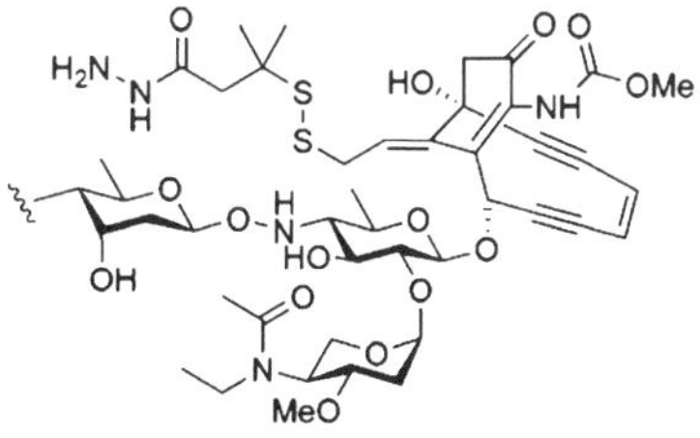

125

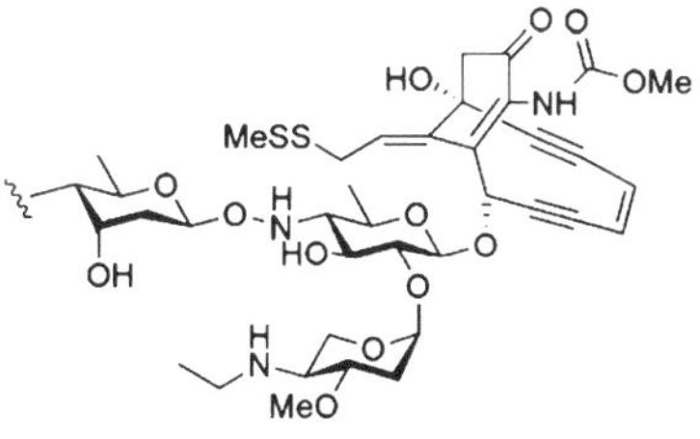

126

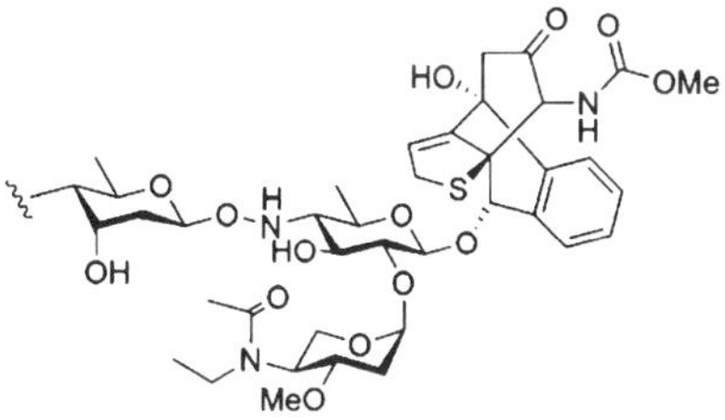

127

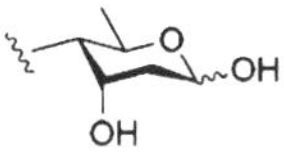

128

Figure 25. Alternative acceptors for CalG1-catalyzed sugar exchange reactions.

native sugar had been replaced by 3-deoxy-α-D-glucose (Figure 24, **114**). To demonstrate the potential utility of sugar exchange, a combination of 10 CalG1 dTDP-sugar donors (Figure 24, **25–29**, **105–110**) and 8 calicheamicin variants (Figure 25, **122–128**) was subsequently employed for the production of a CLM library exceeding 70 members (33). The same study, revealed a representative aglycon exchange reaction in which TDP-6-azido-glucose **108** (generated from the reverse reaction catalyzed by GtfE with azido-glucoside **129**) served as a CalG1 donor in a tandem one-pot reaction containing calicheamicin aglycone **111**—the net result being the transfer

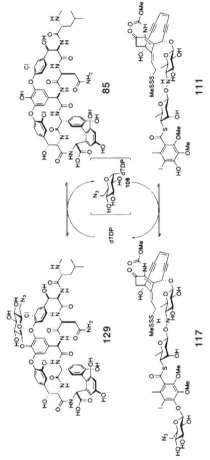

Figure 26. Aglycon exchange reaction catalyzed by CalG1.

of an "unnatural" sugar (6-azido-glucose) between vancomycin and calicheamicin scaffolds (Figure 26).

The discovery of the reversibility of GT reactions also provided a
convenient method to bypass the prohibitive requirement of synthesizing
exotic native sugar nucleotides for in vitro GT assays. For example, the
activity of CalG4 was determined via assays for the reverse reaction
(specifically, the TDP-dependent removal of the amino-pentose moiety
from aglycones **122–124** and **125**) and aglycon exchange (wherein the
excised amino-pentose was transferred to exogenous aglycon acceptor
111) (33). Similar studies have now been successfully conducted on the
avermectin GT AveB1 (Section III.A.4) (35), erythromycin GT EryBV
(Section III.A.2) (30), vancomycin GTs GtfB and GtfD (Section III.B.1)
(33), calicheamicin GTs CalG2 and CalG3 (C. Zhang and J. S. Thorson,
unpublished), and polyene GTs NysD1 and AmpD1 (C. Zhang and J. S.
Thorson, unpublished).

E.   MACROLACTAMS

1. Vicenistatin

Vicenistatin (Figure 27, **130**) is an antitumor polyketide glycoside
produced by *Streptomyces halstedii* HC-34 (134). The GT responsible for
the transfer of the amino sugar vicenisamine from **131** to the 20-membered
macro-lactam aglycone **132** was recently identified as VinC following

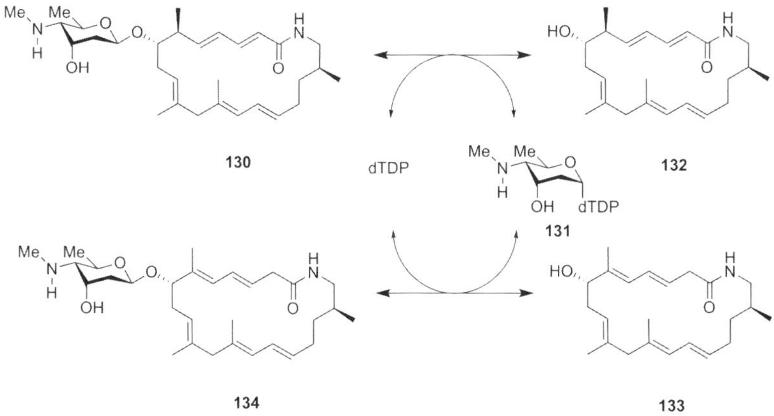

Figure 27. Reaction catalyzed by VinC.

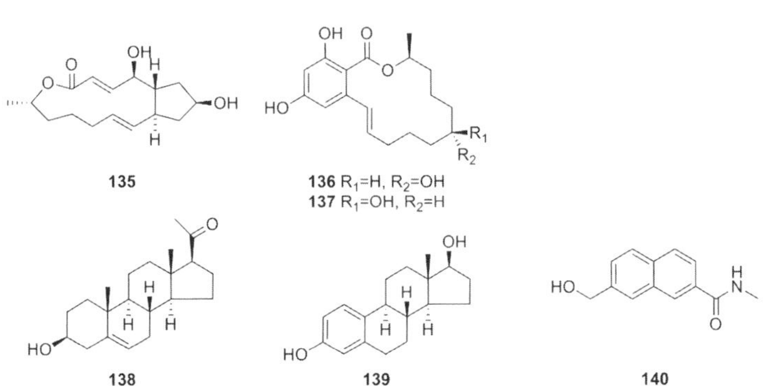

Figure 28.   Unnatural acceptors for the VinC-catalyzed transfer of TDP-vicenisamine **119**.

characterization of the vicenistatin biosynthetic gene cluster (135). Minami et al. also provided evidence that VinC could catalyze the reverse glycosylation reaction from vicenistatin and dTDP and that in situ–generated dTDP-vicenisamine **131** could be transferred in situ to an alternative aglycone, neovicenilactam (**133**) (136). Subsequently, this "aglycon switch" was also performed with a diverse panel of aglycons (Figure 28) that were identified as acceptors for the forward reaction, which included brefaldin A (**135**), α- and β-zearalenol (**136** and **137**), β-estradiol (**138**), and pregnenolone (**139**) (137). Based upon this work, the authors concluded that hydrogen bonding interactions between either the amide of the natural macrolactam or the oxygen functionalities of the unnatural aglycons were most important determinants of recognition, along with overall molecular size. Consistent with this postulation, the designed unnatural aglycon **140**, containing an appropriately spaced alcohol and amide, was a good substrate for VinC (137). VinC was recently shown to catalyze the glycosyl transfer from *both* the α- and β-anomers of several unnatural dTDP-2-deoxy-D-sugars, yielding the respective β- and α-glycoside, albeit at rates several orders of magnitude slower than that with the natural donor (138). While there now exist many examples of GTs capable of using both α-D- and β-L-configured NDP-sugars, this is the first report of a GT able to utilize *both* anomers of a single NDP-sugar, and it remains to be seen whether other GTs are capable of the same feat. To explain activity with β-anomers of dTDP-D-sugars, the authors suggest that a conformational change of the hexose part of the donor allows the NDP moiety to occupy a pseudoaxial position, which consequently allows catalysis to proceed via the oxocarbenium transition state.

F. AROMATIC POLYKETIDES

1. Anthracyclines

Anthracyclines are potent antitumor microbial antibiotics, exemplified by doxorubicin (adriamycin) (Figure 29, **141**), daunorubicin (daunomycin) (Figure 29, **142**), and the related aclarubicins (aclacinomycin A) (Figure 29, **143**). Anthracyclines remain among the top cytotoxic anticancer agents employed today—for example, doxorubicin remains one of the most effective clinical agents available for the treatment of solid tumors (139, 140). Yet, while more than 2000 anthracycline derivatives have been synthesized and studied to date (141), the precise mechanism of anthracycline biological action in vivo is still poorly understood. Like other DNA-intercalating agents, anthracyclines are topoisomerase II inhibitors and lead to DNA damage (142, 143). However, more recent studies have revealed anthracyclines to

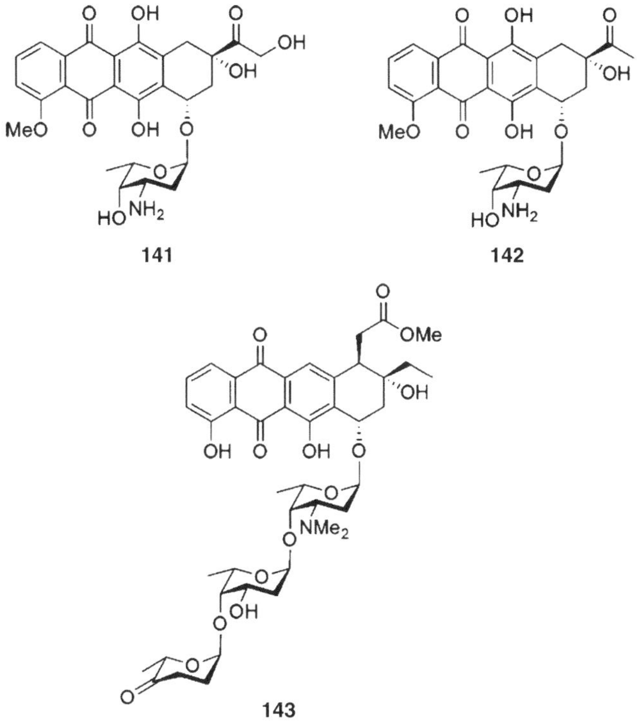

Figure 29. Natural and engineered anthracyclines.

also inhibit topoisomerase I in a concentration-dependent manner and at higher concentrations induce the formation of covalent drug–DNA–enzyme ternary complexes (142). Although drug intercalation may have a role in the mechanism of enzyme inhibition, external interactions involving the sugar residue(s) and the cyclohexene ring are expected to critically influence formation of ternary complexes. Moreover, while the sugars are clearly important for drug–DNA interactions (144), sugar alterations have attributed to reduced toxicity (as found with the semisynthetic C-4′-doxorubicin sugar epimer epirubicin) (143), the broadening of antitumor spectrum (as illustrated by the semisynthetic disaccharide derivative MEN-10755) (145), and the reduction of multidrug resistance (MDR) efflux (as with **143** and C-3′-*N,N*-dimethyl-**141**) (146). Finally, anthracycline deglycosidation (hydrolysis of the sugar) is a significant contributing factor to the cardiac toxicity of anthracyclines (147, 148), and the carbohydrate–DNA recognition has also been implicated in the catalytic redox-cycling activity of anthracyclines, which leads to free-radical-based cellular damage (149, 150).

Recently, the in vitro study of several anthracycline GTs have been described. The biosynthetic gene cluster for aclacinomycin A (**143**) contains only two GTs, AknS and AknK (151), implicating that one these GTs may catalyze the addition of two sugar moieties to the aclacinomycin A aglycone (Figure 30, **144**). Accordingly, early in vitro studies revealed AknK to be responsible for transfer of the second and third sugar to the trisaccharide chain (152). In this study, AknK catalyzed the addition of 2-deoxy-L-fucose to rhodinosaminyl aklavinone **144** with a k_{cat} of 66 min^{-1} and K_m for dTDP-2-deoxy-L-fucose and **144** of 149 and 109 µM, respectively. In addition, the enzyme accepted an alternative donor, dTDP-L-daunosamine, albeit with a 270-fold reduced catalytic efficiency (k_{cat}/K_m) (Figure 30). Furthermore, AknK could also transfer 2-deoxy-L-fucose to an alternative aglycon, idarubicin, with k_{cat} of 5.6 min^{-1} and K_m values for dTDP-2-deoxy-L-fucose and aglycone of 562 and 137 µM. A notable increase in K_m (about fourfold) for the native NDP-sugar was observed in the presence of the unnatural aglycon idarubicin, consistent with a similar interdependence of donor and acceptor specificity found with a number of other GTs. Adriamycin (**141**) and daunomycin (**142**) were also acceptors with dTDP-2-deoxy-L-fucose as donor, while daunosamine was only transferred to idarubicin. In contrast, epirubicin was not a substrate. AknK was shown to also transfer a second 2-deoxy-L-fucose to 2-deoxyfucosyl-rhodinosaminyl-aklavinone **145** at a rate at least about three orders of magnitude slower than the first deoxyfucose attachment (Figure 30).

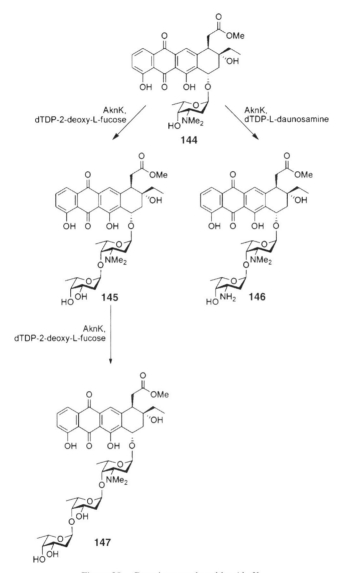

Figure 30. Reactions catalyzed by AknK.

The discovery that DesVII requires an activating partner, DesVIII, for full activity (75) (see also Section III.A.1) and the presence of a *desVIII* homolog (*aknT*) in the aclacinomycin biosynthetic gene cluster raised the possibility that AknS also required such an activating partner. Subsequently, the k_{cat} of recombinant AknS was stimulated 40-fold for the transfer of the nonnatural moiety 2-deoxy-L-fucose to the aglycon aklavinone (Figure 31, **148**) in the presence of recombinant AknT (153). Yet, this transfer still occurred with poor efficiency, with a k_{cat} of 0.22 min^{-1} and K_m values for dTDP-2-deoxy-L-fucose and aglycone **148** of 290 and 3.5 μM, respectively. In vitro reconstitution of AknS/AknT activity also allowed a limited examination of the donor and acceptor specificity. The alternative donor, dTDP-4-amino-2-deoxy-L-rhamnose displayed a k_{cat} similar to that of dTDP-2-deoxy-L-fucose using aklavinone **148** as the acceptor while 2-deoxy-L-fucose could also be transferred to the alternative scaffold, ε-rhodomycinone **149** (Figure 31). The sequential action of AknS/AknT and AknK has also been exploited in vitro wherein AknK transferred 2-deoxy-L-fucose or daunosamine (4-amino-2-deoxyrhamnose) to two alternate 2-deoxyfucosyl-aklavinone aglycones **150** and **151** producing the novel glycosides **152–155** (Figure 31) (153). Recently, the chemical synthesis of dTDP-L-rhodosamine (from dTDP-L-daunosamine) allowed the complete assignment of the true in vivo role of AknS/AknT (154). AknS/AknT was shown to transfer dTDP-rhodosamine to the aglycon aklavinone (**148**) with a k_{cat} of 9.6 min^{-1} and K_m values for the donor and aglycon of 280 and 5.7 μM, respectively (Figure 32). While the activity with the native substrate was nearly 50-fold higher than with the nonnatural donor, interestingly the K_m for aglycon or donor remains largely unchanged when either donor was used. Intriguingly, binding measurements between aklavinone and AknS/AknT originally suggested that the activating effect of AknT on glycosyl transfer correlated with tighter binding of the aglycone substrate, as judged by determination of the K_d by fluorescence (153). Yet, recent determination of the kinetic parameters of the transfer of the natural donor to aklavinone by AknS in the presence and absence of AknT revealed that the K_m for both aglycon and donor remain largely unchanged by the presence of AknT (154). However, improved k_{cat} could well be reflected by a concomitant improved K_d without a necessary change in K_m, since the K_m for a bi–bi reaction mechanism is complex. Thus, AknT may allow AknS to access an improved transition state (154). Following glycosylation by AknS/AknT, AknK sequentially transfers dTDP-2-deoxy-L-fucose and then dTDP-rhodinose to afford aclacinomycin A (**143**) after oxidation (Figure 32).

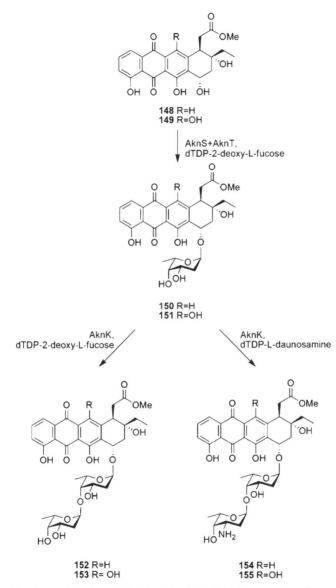

Figure 31. Sequential action of AknS/AknT and AknK in the generation of anthracycline glycosides.

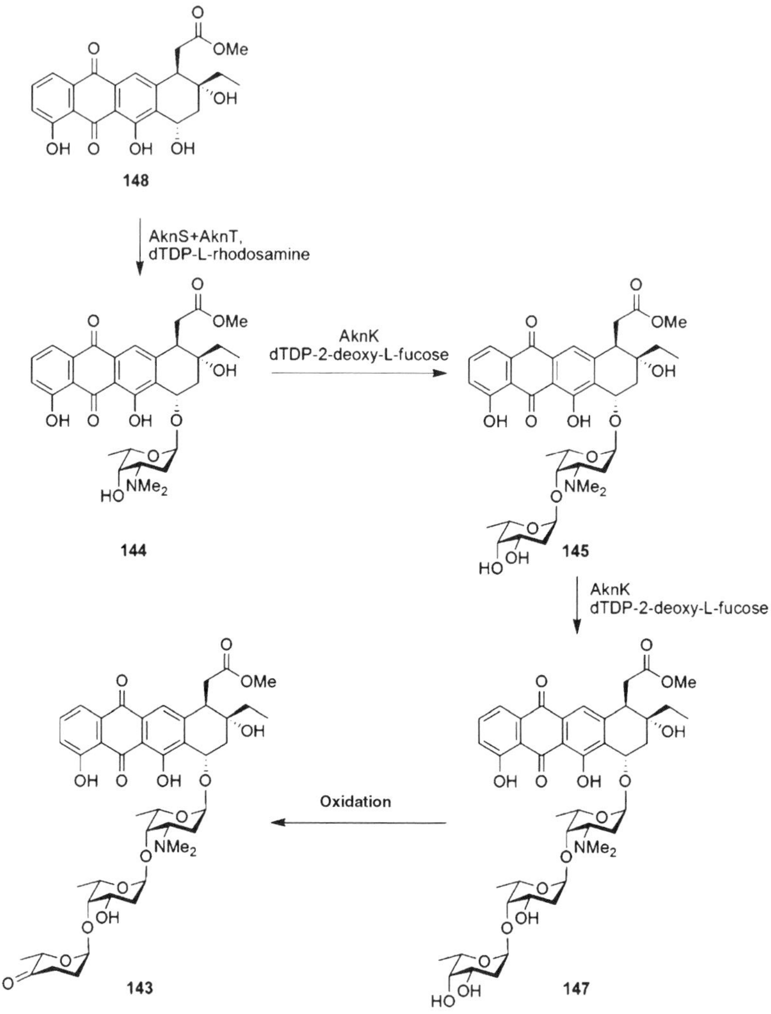

Figure 32.   Biosynthetic pathway of aclacinomycin A.

The GT gene *araGT* required for the rhamnosylation of aranciamycin in *Streptomyces echinatus* has also been cloned and overexpressed and the corresponding recombinant AraGT purified and characterized (155). Based upon a mass spectrometry assay, AraGT catalyzed the transfer of rhamnose

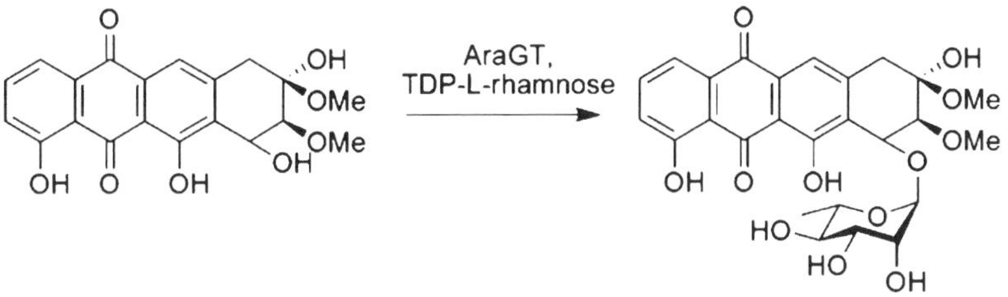

156   **157**

Figure 33.   Reaction catalyzed by AraGT with the unnatural acceptor **156**.

from dTDP-L-rhamnose to the aranciamycin analog **156** (Figure 33) in this preliminary study.

IV.   Engineering Glycosyltransferases

While some natural product GTs are remarkably flexible with regard to donor/acceptor, none of the GTs studied to date are sufficiently promiscuous for use in a truly combinatorial capacity. Compounding this problem is the apparent codependence of donor/acceptor specificity observed in many natural product GTs. Recent efforts to begin to circumvent these limitations, with the ultimate goal of creating truly combinatorial GTs, can be broadly classified into rational (structure-based and sequence-guided) enzyme redesign or directed evolution.

A.   RATIONAL (SEQUENCE-GUIDED) DESIGN

Primary amino acid sequences are the key to the CAZY classification of GTs (http://afmb.cnrs-mrs.fr/~cazy/CAZY/index.html) and, as previously mentioned, the CAZY database contains 23,836 sequences predicted or known GT sequences that are divided into 89 families on the basis of amino acid sequence similarity and mechanistic similarities (where known). Such sequence alignments can also provide a basis for retooling GT catalysis as exemplified by a landmark study of Hoffmeister et al. aimed at identifying the residues responsible for donor selectivity in the urdamycin GTs. From in vivo studies, UrdGT1c was known to catalyze the transfer of dNDP-L-rhodinose to **156** (Figure 34) to provide urdamycin G **157** while UrdGT1b was responsible for the transfer of dNDP-D-olivose to

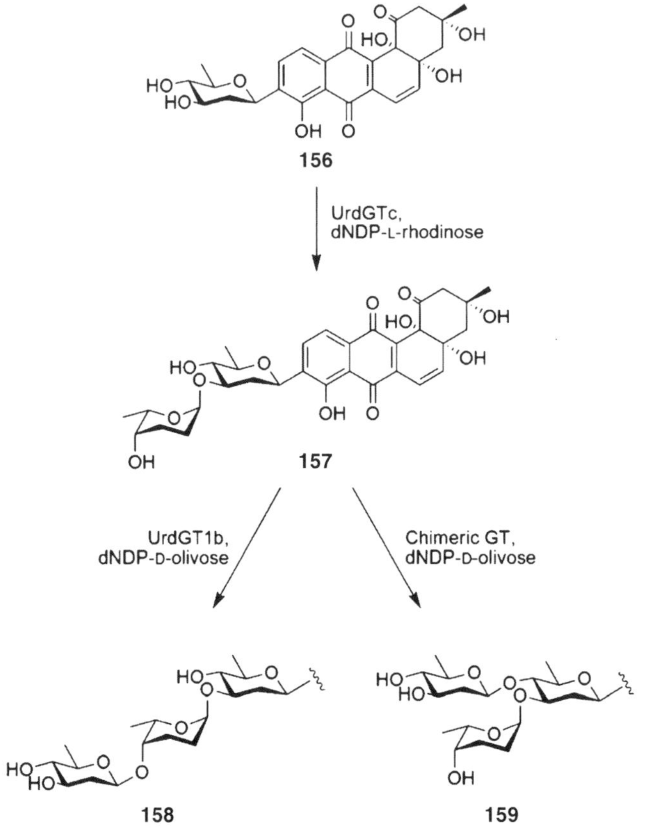

Figure 34.   Reactions catalyzed by natural and chimeric urdamycin GTs.

157 to furnish urdamycin A (**158**) (156, 157). Despite differences in substrate and linkage specificity, these two enzymes share 91% amino acid identity. The in vivo analysis of 10 UrdGT1b/1c chimeras revealed a single region, spanning amino acids 52–82, conferred donor and acceptor specificity (158). A subsequent comprehensive mutagenesis study of this region in UrdGT1b delineated specific residues capable of conferring specificity alterations (159). Interestingly, specific activities were supported by multiple sets of amino acid mutations but not achieved through single-amino-acid changes from either wild-type enzyme. Yet, some of the mutant UrdGTs differed by only a single mutation and proved sufficient to generate novel activities, such as generation of **159** (Figure 34). Intriguingly, the sequence element identified in this study corresponds to a region that is often highly variable among other natural product GTs (48, 51, 54).

In another example, multiple alignments of glucosyl- and galactosyltransferases from a range of sources identified a C-terminal position in the UDPGT signature motif that may determine C4-epimer specificity (160). Consequently, introduction of the H374Q mutation into a plant UDP-galactose–anthocyanin GT improved k_{cat}/K_m for UDP-glucose approximately 26-fold, while the specificity constant toward UDP-galactose was unchanged. However, the "reverse" mutation (Q382H) in a UDP-glucose–flavonoid GT did not invoke a significant change in specificity between UDP-glucose and UDP-galactose and actually reduced overall activity several-fold (160). Thus, consistent with the UrdGT study (159), even very subtle specificity alterations (e.g., differentiating simple C4-epimers) are typically dictated by more than residue. Yet, in contrast to the UrdGT study, swapping larger sequence elements of plant GTs failed to achieve the desired donor specificity changes. For example, the potato glycosyltransferase StSGT galactosylates various steroidal acceptors, while the homologous (75% sequence homology) SaGT4a from *Solanum aculeatissimum* transfers glucose to an identical acceptor array. Variable-sequence elements in the C-terminal domain were swapped between the two enzymes but none of the resulting chimeras displayed changes in donor specificity (161).

B.   RATIONAL (STRUCTURE-GUIDED) DESIGN

Structure-based redesign requires detailed structural and mechanistic information. Thus, given the wealth of structural/mechanistic information for GT-A GTs, it is not surprising that the greatest effort has been focused upon oligosaccharide-forming GT-A GTs. These studies have been reviewed extensively (162, 163), and while they are not directly relevant to classical natural products, they are demonstrative of structure-based GT specificity redesign. For example, understanding substrate specificity in the bovine β-1,4-galactosyltransferase led to a rational mutant (R228K) which favored glucosyl transfer by 90-fold (164). Another mutation, Y289L, expanded the substrate tolerance to include nonnatural 2-keto sugar donors (165, 166).

Given the more recent emergence of GT-B structural and detailed mechanistic studies, successful examples of GT-B fold engineering are far less prevalent. Structure-guided mutagenesis was recently used to alter the

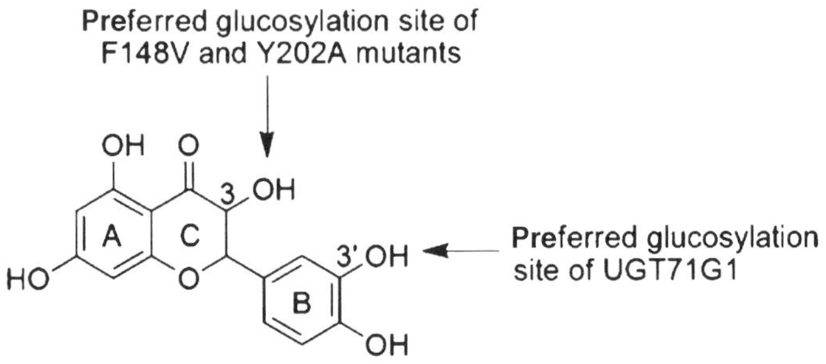

**Preferred glucosylation site of
F148V and Y202A mutants**

OH O

A C 3 OH

HO O B 3' OH ← **Preferred glucosylation
site of UGT71G1**

OH

59

Figure 35. Regioselectivity of a wild-type and engineered plant GT toward quercetin **59**.

regioselectivity of the legume flavonoid GT from *Medicago truncatula* (167), which can transfer glucose to each of the five hydroxyls of quercetin (Figure 35, **59**) with the 3'-*O*-glucoside (B ring) as the major product. In this study, the systematic mutagenesis of residues that form the acceptor binding pocket led to the identification of two mutations (F148V and Y202A) which shifted glycosylation toward the 3-position (C ring) such that the 3-*O*-glucoside was the major product while maintaining good overall efficiency (Figure 35).

C. DIRECTED EVOLUTION OF GLYCOSYLTRANSFERASES

The lack of high-throughput GT assays has hampered GT-directed evolution experiments, which typically require screening at least a few thousand variants. Recently, a truly high-throughput method was developed for screening oligosaccharide-forming GT-A sialyltransferases (STs) (168). While not directly relevant to classical natural product GTs, this ingenious assay was based upon trapping a cell-permeable fluorescent acceptor via intracellular ST-catalyzed sialylation (attachment of a charged sugar). Expression of mutant STs in an *E. coli* strain engineered to produce CMP-sialic acid (the sugar donor) followed by growth in the presence of bodipy-lactose (the fluorescent acceptor) and fluorescent-activated cell sorting (FACS) led to the identification of a mutant (F91Y) which displayed a 153-fold increase in k_{cat}/K_m toward the bodipy-lactose (versus lactose) and a 367-fold improvement toward the bodipy-galactose (versus galactose). Transfer of CMP-sialic to either fluorescently tagged acceptor could not be detected by the wild-type enzyme. Determination of the mutant ST crystal structure revealed the formation a new hydrophobic pocket, created by the F91Y substitution, to accommodate the fluorescent tag. As the first reported example of GT-directed evolution, this notable study opened the door to a vast array of new GT-associated opportunities. Likewise, although the screen employed required charged donors accessible via natural in vivo pathways and noncytotoxic fluorescent acceptors, it stands as a powerful new tool in GT-directed evolution.

The promiscuity of a GT-B fold natural product GT has also been recently expanded via directed evolution (169). As previously noted (Section III.A.3), the oleandomycin GT OleD glucosylates a range of small aromatic phenolics, including the fluorescent umbelliferone, 7-hydroxy-4-methylcoumarin (Figure 36, **51**). Based upon the known fluorescent properties of coumarins and substituted coumarins (170, 171), the elimination of fluorescence upon umbelliferone 7-glycosylation was used to screen a library of OleD random mutants created by an error-prone polymerase chain reaction (PCR). Upon screening approximately 1000 OleD variants, three improved GTs were identified from which the subsequent combination of functional mutations led to the triple mutant P67T–S132F–A242V. This triple mutant was 60-fold more efficient toward the unnatural acceptor umbelliferone and UDP-glucose and this increase in *proficiency* translated to an impressive expansion of *promiscuity* toward a panel of donors and acceptors. Specifically, the evolved triple mutant accepted 15 of 22 NDP donors tested (Figure 36, **47, 69, 160–171**), 12 of which were nondetectable substrates for the WT enzyme. Moreover, the evolved enzyme displayed improvements toward several acceptors tested, including flavonoids (**58**), coumarins (**54** and **172**), isoflavones (**173** and **174**), and aminocoumarin (**98**). Interestingly, the functional OleD mutations mapped to the active site of the enzyme. For example, the most effective mutation in terms of acceptor improvements, P67T, is located in the N3 hypervariable loop region which follows β-sheet 3 in the N-terminal domain of the GT-B fold discussed in the previously described UrdGT engineering efforts (see Section IV.A). Analysis of the crystal structure of OleD (49) and comparison to the plant flavonoid VvGT1 (55) structure also allowed some interpretation for the other two functional mutations in the evolved enzyme. Ser-132 likely forms a hydrogen bond to the C6-OH of the donor while Ala-242 is likely involved in binding the diphosphate moiety of the NDP donor and is immediately adjacent to Ser-241, a highly conserved residue in GT-B fold GTs.

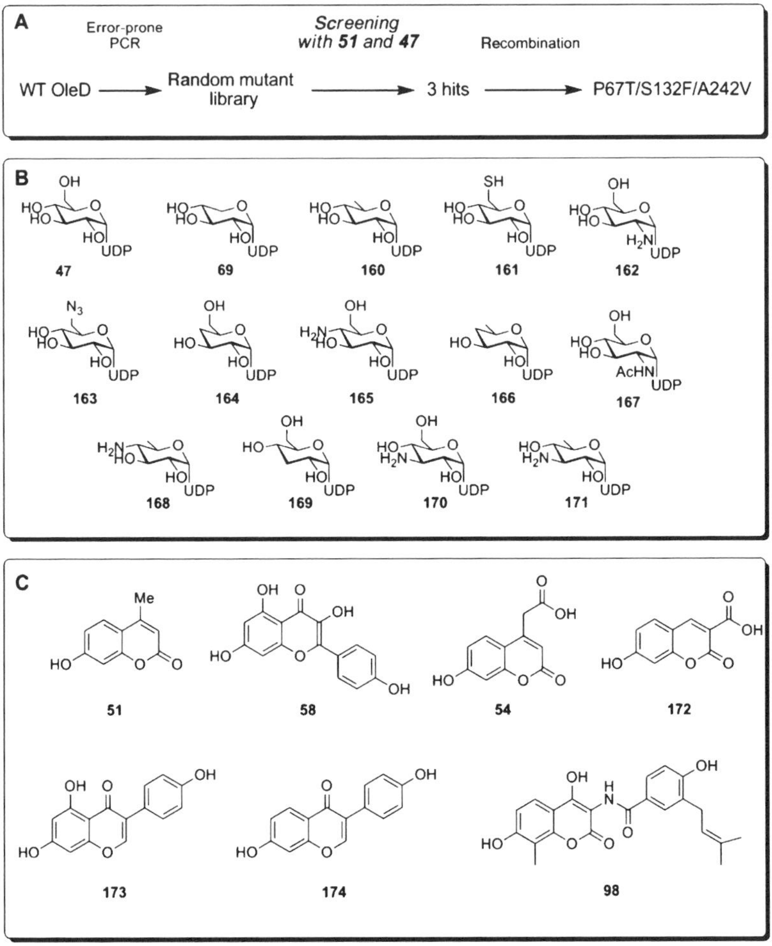

Figure 36.   Directed evolution of a natural product GT-B fold GT, OleD. (A) Simple screen for glycosyl transfer activity. (B) NDP donor specificity of evolved enzyme. (C) Acceptor specificity of evolved enzyme.

V.   Conclusions

Natural product GTs have proved incredibly useful tools for the synthesis of novel glycosides for the discovery of new and improved drugs. Despite the huge variation of substrate structures accepted by natural product GTs, most GTs fall into one of two fold types, GT-A and

GT-B. Most natural product GTs belong to the latter class and operate upon an architecturally diverse array of therapeutically important natural products, including aminoglycosides, aminocoumarins, enediynes, glycopeptides, macrolides, macrolactams, and aromatic polyketides. The combination of chemical and chemoenzymatic NDP-sugar synthesis has allowed investigation of the substrate specificity of these GTs and has led to the realization that many natural product GTs display broad substrate promiscuity. Moreover, the recent exploitation of the reversibility of GT catalysis has greatly elevated the potential impact of GTs in natural product glycodiversification. Recently devised screening strategies have further aided the creation of GTs with novel specificities by directed evolution. In the context of many fundamental advances highlighted within this review, there is no doubt that this is a very exciting time for natural product–associated glycocatalysts and the future of the field holds significant promise.

References

1. Butler, M. S. (2004) The role of natural product chemistry in drug discovery, *J. Nat. Prod.* *67*, 2141–2153.

2. Butler, M. S. (2005) Natural products to drugs: natural product derived compounds in clinical trials, *Nat. Prod. Rep. 22*, 162–195.

3. Butler, M. S., and Buss, A. D. (2006) Natural products—The future scaffolds for novel antibiotics? *Biochem. Pharmacol. 71*, 919–929.

4. Gullo, V. P., McAlpine, J., Lam, K. S., Baker, D., and Petersen, F. (2006) Drug discovery from natural products, *J. Ind. Microbiol. Biotechnol. 33*, 523–531.

5. Weymouth-Wilson, A. C. (1997) The role of carbohydrates in biologically active natural products, *Nat. Prod. Rep. 14*, 99–110.

6. Griffith, B. R., Langenhan, J. M., and Thorson, J. S. (2005) "Sweetening" natural products via glycorandomization, *Curr. Opin. Biotechnol. 16*, 622–630.

7. Blanchard, S., and Thorson, J. S. (2006) Enzymatic tools for engineering natural product glycosylation, *Curr. Opin. Chem. Biol. 10*, 263–271.

8. Ge, M., Chen, Z., Onishi, H. R., Kohler, J., Silver, L. L., Kerns, R., Fukuzawa, S., Thompson, C., and Kahne, D. (1999) Vancomycin derivatives that inhibit peptidoglycan biosynthesis without binding D-Ala-D-Ala, *Science 284*, 507–511.

9. Solenberg, P. J., Matsushima, P., Stack, D. R., Wilkie, S. C., Thompson, R. C., and Baltz, R. H. (1997) Production of hybrid glycopeptide antibiotics in vitro and in *Streptomyces toyocaensis*, *Chem. Biol. 4*, 195–202.

10. Kahne, D., Leimkuhler, C., Lu, W., and Walsh, C. (2005) Glycopeptide and lipoglycopeptide antibiotics, *Chem. Rev. 105*, 425–448.

11. Ahmed, A., Peters, N. R., Fitzgerald, M. K., Watson, J. A., Jr., Hoffmann, F. M., and Thorson, J. S. (2006) Colchicine glycorandomization influences cytotoxicity and mechanism of action, *J. Am. Chem. Soc. 128*, 14224–14225.

12. Langenhan, J. M., Peters, N. R., Guzei, I. A., Hoffmann, F. M., and Thorson, J. S. (2005) Enhancing the anticancer properties of cardiac glycosides by neoglycorandomization, *Proc. Natl. Acad. Sci. USA 102*, 12305–12310.

13. Zhu, L., Cao, X., Chen, W., Zhang, G., Sun, D., and Wang, P. G. (2005) Syntheses and biological activities of daunorubicin analogs with uncommon sugars, *Bioorg. Med. Chem. 13*, 6381–6387.

14. Wei, L., Wei, G., Zhang, H., Wang, P. G., and Du, Y. (2005) Synthesis of new, potent avermectin-like insecticidal agents, *Carbohydr. Res. 340*, 1583–1590.

15. Zhang, G., Fang, L., Zhu, L., Zhong, Y., Wang, P. G., and Sun, D. (2006) Syntheses and biological activities of 3′-azido disaccharide analogues of daunorubicin against drug-resistant leukemia, *J. Med. Chem. 49*, 1792–1799.

16. Shan, M., and O'Doherty, G. A. (2006) De novo asymmetric syntheses of SL0101 and its analogues via a palladium-catalyzed glycosylation, *Org. Lett. 8*, 5149–5152.

17. Guo, H., and O'Doherty, G. A. (2007) De novo asymmetric synthesis of the anthrax tetrasaccharide by a palladium-catalyzed glycosylation reaction, *Angew. Chem. Int. Ed. Engl. 46*, 5206–5208.

18. Zhou, M., and O'Doherty, G. A. (2007) De novo approach to 2-deoxy-β-glycosides: Asymmetric syntheses of digoxose and digitoxin, *J. Org. Chem. 72*, 2485–2493.

19. Langenhan, J. M., Griffith, B. R., and Thorson, J. S. (2005) Neoglycorandomization and chemoenzymatic glycorandomization: Two complementary tools for natural product diversification, *J. Nat. Prod. 68*, 1696–1711.

20. Thibodeaux, C. J., Melancon, C. E., and Liu, H. W. (2007) Unusual sugar biosynthesis and natural product glycodiversification, *Nature 446*, 1008–1016.

21. Salas, J. A., and Mendez, C. (2007) Engineering the glycosylation of natural products in *actinomycetes*, *Trends. Microbiol. 15*, 219–232.

22. Rodriguez, L., Aguirrezabalaga, I., Allende, N., Brana, A. F., Mendez, C., and Salas, J. A. (2002) Engineering deoxysugar biosynthetic pathways from antibiotic-producing microorganisms. A tool to produce novel glycosylated bioactive compounds, *Chem. Biol. 9*, 721–729.

23. Perez, M., Lombo, F., Zhu, L., Gibson, M., Brana, A. F., Rohr, J., Salas, J. A., and Mendez, C. (2005) Combining sugar biosynthesis genes for the generation of L- and D-amicetose and formation of two novel antitumor tetracenomycins, *Chem. Commun. (Camb).* 1604–1606.

24. Lombo, F., Gibson, M., Greenwell, L., Brana, A. F., Rohr, J., Salas, J. A., and Mendez, C. (2004) Engineering biosynthetic pathways for deoxysugars: Branched-chain sugar pathways and derivatives from the antitumor tetracenomycin, *Chem. Biol. 11*, 1709–1718.

25. Perez, M., Lombo, F., Baig, I., Brana, A. F., Rohr, J., Salas, J. A., and Mendez, C. (2006) Combinatorial biosynthesis of antitumor deoxysugar pathways in *Streptomyces griseus*: Reconstitution of "unnatural natural gene clusters" for the biosynthesis of four 2,6-D-dideoxyhexoses, *Appl. Environ. Microbiol. 72*, 6644–6652.

26. Blanco, G., Patallo, E. P., Brana, A. F., Trefzer, A., Bechthold, A., Rohr, J., Mendez, C., and Salas, J. A. (2001) Identification of a sugar flexible glycosyltransferase from *Streptomyces olivaceus*, the producer of the antitumor polyketide elloramycin, *Chem. Biol. 8*, 253–263.

27. Salas, A. P., Zhu, L., Sanchez, C., Brana, A. F., Rohr, J., Mendez, C., and Salas, J. A. (2005) Deciphering the late steps in the biosynthesis of the anti-tumour indolocarbazole staurosporine: Sugar donor substrate flexibility of the StaG glycosyltransferase, *Mol. Microbiol. 58*, 17–27.

28. Sanchez, C., Mendez, C., and Salas, J. A. (2006) Engineering biosynthetic pathways to generate antitumor indolocarbazole derivatives, *J. Ind. Microbiol. Biotechnol. 33*, 560–568.

29. Albermann, C., Soriano, A., Jiang, J., Vollmer, H., Biggins, J. B., Barton, W. A., Lesniak, J., Nikolov, D. B., and Thorson, J. S. (2003) Substrate specificity of NovM: Implications for novobiocin biosynthesis and glycorandomization, *Org. Lett. 5*, 933–936.

30. Zhang, C., Fu, Q., Albermann, C., Li, L., and Thorson, J. S. (2007) The *in vitro* characterization of the erythronolide mycarosyltransferase EryBV and its utility in macrolide diversification, *Chembiochem 8*, 385–390.

31. Fu, X., Albermann, C., Jiang, J., Liao, J., Zhang, C., and Thorson, J. S. (2003) Antibiotic optimization via in *vitro* glycorandomization, *Nat. Biotechnol. 21*, 1467–1469.

32. Fu, X., Albermann, C., Zhang, C., and Thorson, J. S. (2005) Diversifying vancomycin via chemoenzymatic strategies, *Org. Lett. 7*, 1513–1515.

33. Zhang, C., Griffith, B. R., Fu, Q., Albermann, C., Fu, X., Lee, I. K., Li, L., and Thorson, J. S. (2006) Exploiting the reversibility of natural product glycosyltransferase-catalyzed reactions, *Science 313*, 1291–1294.

34. Zhang, C., Albermann, C., Fu, X., Peters, N. R., Chisholm, J. D., Zhang, G., Gilbert, E. J., Wang, P. G., Van Vranken, D. L., and Thorson, J. S. (2006) RebG- and RebM-catalyzed indolocarbazole diversification, *Chembiochem 7*, 795–804.

35. Zhang, C., Albermann, C., Fu, X., and Thorson, J. S. (2006) The *in vitro* characterization of the iterative avermectin glycosyltransferase AveBI reveals reaction reversibility and sugar nucleotide flexibility, *J. Am. Chem. Soc. 128*, 16420–16421.

36. Campbell, J. A., Davies, G. J., Bulone, V., and Henrissat, B. (1997) A classification of nucleotide-diphospho-sugar glycosyltransferases based on amino acid sequence simi-larities, *Biochem. J. 326* (Pt. 3), 929–939.

37. Sinnott, M. L. (1990) Catalytic mechanisms of enzymatic glycosyl transfer, *Chem. Rev. 90*, 1171–1202.

38. Lairson, L. L., Henrissat, B., Davies, G. J., and Withers, S. G. (2008) Glycosyltransferase: Structures, functions, and mechanisms, *Annu. Rev. Biochem. 77*, 521–555.

39. Martinez-Fleites, C., Proctor, M., Roberts, S., Bolam, D. N., Gilbert, H. J., and Davies, G. J. (2006) Insights into the synthesis of lipopolysaccharide and antibiotics through the structures of two retaining glycosyltransferases from family GT4, *Chem. Biol. 13*, 1143–1152.

40. Unligil, U. M., and Rini, J. M. (2000) Glycosyltransferase structure and mechanism, *Curr. Opin. Struct. Biol. 10*, 510–517.

41. Bourne, Y., and Henrissat, B. (2001) Glycoside hydrolases and glycosyltransferases: Families and functional modules, *Curr. Opin. Struct. Biol. 11*, 593–600.

42. Coutinho, P. M., Deleury, E., Davies, G. J., and Henrissat, B. (2003) An evolving hierarchical family classification for glycosyltransferases, *J. Mol. Biol. 328*, 307–317.

43. Persson, K., Ly, H. D., Dieckelmann, M., Wakarchuk, W. W., Withers, S. G., and Strynadka, N. C. (2001) Crystal structure of the retaining galactosyltransferase LgtC from *Neisseria meningitidis* in complex with donor and acceptor sugar analogs, *Nat. Struct. Biol. 8*, 166–175.

44. Flint, J., Taylor, E., Yang, M., Bolam, D. N., Tailford, L. E., Martinez-Fleites, C., Dodson, E. J., Davis, B. G., Gilbert, H. J., and Davies, G. J. (2005) Structural dissection and high-throughput screening of mannosylglycerate synthase, *Nat. Struct. Mol. Biol. 12*, 608–614.

45. Breton, C., Bettler, E., Joziasse, D. H., Geremia, R. A., and Imberty, A. (1998) Sequence-function relationships of prokaryotic and eukaryotic galactosyltransferases, *J. Biochem. 123*, 1000–1009.

46. Breton, C., and Imberty, A. (1999) Structure/function studies of glycosyltransferases, *Curr. Opin. Struct. Biol. 9*, 563–571.

47. Murray, B. W., Takayama, S., Schultz, J., and Wong, C. H. (1996) Mechanism and specificity of human α-1,3-fucosyltransferase V, *Biochemistry 35*, 11183–11195.

48. Mulichak, A. M., Lu, W., Losey, H. C., Walsh, C. T., and Garavito, R. M. (2004) Crystal structure of vancosaminyltransferase GtfD from the vancomycin biosynthetic pathway: Interactions with acceptor and nucleotide ligands, *Biochemistry 43*, 5170–5180.

49. Bolam, D. N., Roberts, S., Proctor, M. R., Turkenburg, J. P., Dodson, E. J., Martinez-Fleites, C., Yang, M., Davis, B. G., Davies, G. J., and Gilbert, H. J. (2007) The crystal structure of two macrolide glycosyltransferases provides a blueprint for host cell antibiotic immunity, *Proc. Natl. Acad. Sci. USA 104*, 5336–5341.

50. Hu, Y., and Walker, S. (2002) Remarkable structural similarities between diverse glycosyltransferases, *Chem. Biol. 9*, 1287–1296.

51. Mulichak, A. M., Losey, H. C., Walsh, C. T., and Garavito, R. M. (2001) Structure of the UDP-glucosyltransferase GtfB that modifies the heptapeptide aglycone in the biosynthesis of vancomycin group antibiotics, *Structure 9*, 547–557.

52. Yuan, Y., Barrett, D., Zhang, Y., Kahne, D., Sliz, P., and Walker, S. (2007) Crystal structure of a peptidoglycan glycosyltransferase suggests a model for processive glycan chain synthesis, *Proc. Natl. Acad. Sci. USA 104*, 5348–5353.

53. Lovering, A. L., de Castro, L. H., Lim, D., and Strynadka, N. C. (2007) Structural insight into the transglycosylation step of bacterial cell-wall biosynthesis, *Science 315*, 1402–1405.

54. Mulichak, A. M., Losey, H. C., Lu, W., Wawrzak, Z., Walsh, C. T., and Garavito, R. M. (2003) Structure of the TDP-epi-vancosaminyltransferase GtfA from the chloroeremomycin biosynthetic pathway, *Proc. Natl. Acad. Sci. USA 100*, 9238–9243.

55. Offen, W., Martinez-Fleites, C., Yang, M., Kiat-Lim, E., Davis, B. G., Tarling, C. A., Ford, C. M., Bowles, D. J., and Davies, G. J. (2006) Structure of a flavonoid glucosyltransferase reveals the basis for plant natural product modification, *EMBO J. 25*, 1396–1405.

56. Shao, H., He, X., Achnine, L., Blount, J. W., Dixon, R. A., and Wang, X. (2005) Crystal structures of a multifunctional triterpene/flavonoid glycosyltransferase from *Medicago truncatula*, *Plant Cell*, *17*, 3141–3154.

57. Hans, J., Brandt, W., and Vogt, T. (2004) Site-directed mutagenesis and protein 3D-homology modelling suggest a catalytic mechanism for UDP-glucose-dependent betanidin 5-*O*-glucosyltransferase from *Dorotheanthus bellidiformis*, *Plant J.* *39*, 319–333.

58. Tarbouriech, N., Charnock, S. J., and Davies, G. J. (2001) Three-dimensional structures of the Mn and Mg dTDP complexes of the family GT-2 glycosyltransferase SpsA: a comparison with related NDP-sugar glycosyltransferases, *J. Mol. Biol. 314*, 655–661.

59. Keenleyside, W. J., Clarke, A. J., and Whitfield, C. (2001) Identification of residues involved in catalytic activity of the inverting glycosyl transferase WbbE from *Salmonella enterica serovar borreze*, *J. Bacteriol. 183*, 77–85.

60. Ramakrishnan, B., Boeggeman, E., Ramasamy, V., and Qasba, P. K. (2004) Structure and catalytic cycle of β-1,4-galactosyltransferase, *Curr. Opin. Struct. Biol. 14*, 593–600.

61. Zechel, D. L., and Withers, S. G. (2000) Glycosidase mechanisms: Anatomy of a finely tuned catalyst, *Acc. Chem. Res. 33*, 11–18.

62. Lairson, L. L., and Withers, S. G. (2004) Mechanistic analogies amongst carbohydrate modifying enzymes, *Chem. Commun. (Camb.) 20*, 2243–2248.

63. Gastinel, L. N., Bignon, C., Misra, A. K., Hindsgaul, O., Shaper, J. H., and Joziasse, D. H. (2001) Bovine α,3-galactosyltransferase catalytic domain structure and its relationship with ABO histo-blood group and glycosphingolipid glycosyltransferases, *EMBO J. 20*, 638–649.

64. Gibson, R. P., Turkenburg, J. P., Charnock, S. J., Lloyd, R., and Davies, G. J. (2002) Insights into trehalose synthesis provided by the structure of the retaining glucosyltransferase OtsA, *Chem. Biol. 9*, 1337–1346.

65. Sinnott, M. L., and Jencks, W. P. (1980) Solvolysis of D-glucopyranosyl derivatives in mixtures of ethanol and 2,2,2-trifluoroethanol, *J. Am. Chem. Soc. 102*, 2026–2032.

66. Tvaroska, I. (2004) Molecular modeling insights into the catalytic mechanism of the retaining galactosyltransferase LgtC, *Carbohydr. Res. 339*, 1007–1014.

67. McDaniel, R., Welch, M., and Hutchinson, C. R. (2005) Genetic approaches to polyketide antibiotics. 1, *Chem. Rev. 105*, 543–558.

68. Katz, L., and Ashley, G. W. (2005) Translation and protein synthesis: Macrolides, *Chem. Rev. 105*, 499–528.

69. Bonay, P., Fresno, M., and Alarcon, B. (1997) Megalomicin disrupts lysosomal functions, *J. Cell. Sci. 110* (Pt. 16), 1839–1849.

70. Bonay, P., Duran-Chica, I., Fresno, M., Alarcon, B., and Alcina, A. (1998) Antiparasitic effects of the intra-Golgi transport inhibitor megalomicin, *Antimicrob. Agents Chemother. 42*, 2668–2673.

71. Berisio, R., Schluenzen, F., Harms, J., Bashan, A., Auerbach, T., Baram, D., and Yonath, A. (2003) Structural insight into the role of the ribosomal tunnel in cellular regulation, *Nat. Struct. Biol. 10*, 366–370.

72. Schlunzen, F., Harms, J. M., Franceschi, F., Hansen, H. A., Bartels, H., Zarivach, R., and Yonath, A. (2003) Structural basis for the antibiotic activity of ketolides and azalides, *Structure 11*, 329–338.

73. Xue, Y., Zhao, L., Liu, H. W., and Sherman, D. H. (1998) A gene cluster for macrolide antibiotic biosynthesis in *Streptomyces venezuelae*: Architecture of metabolic diversity, *Proc. Natl. Acad. Sci. USA 95*, 12111–12116.

74. Borisova, S. A., Zhao, L., Melancon, I. C., Kao, C. L., and Liu, H. W. (2004) Characterization of the glycosyltransferase activity of desVII: Analysis of and implications for the biosynthesis of macrolide antibiotics, *J. Am. Chem. Soc. 126*, 6534–6535.

75. Borisova, S. A., Zhang, C., Takahashi, H., Zhang, H., Wong, A. W., Thorson, J. S., and Liu, H. W. (2006) Substrate specificity of the macrolide-glycosylating enzyme pair DesVII/DesVIII: Opportunities, limitations, and mechanistic hypotheses, *Angew. Chem. Int. Ed. Engl. 45*, 2748–2753.

76. Tang, L., and McDaniel, R. (2001) Construction of desosamine containing polyketide libraries using a glycosyltransferase with broad substrate specificity, *Chem. Biol. 8*, 547–555.

77. Zhao, L. S., Ahlert, J., Xue, Y. Q., Thorson, J. S., Sherman, D. H., and Liu, H. W. (1999) Engineering a methymycin/pikromycin-calicheamicin hybrid: Construction of two new macrolides carrying a designed sugar moiety, *J. Am. Chem. Soc. 121*, 9881–9882.

78. Kao, C. L., Borisova, S. A., Kim, H. J., and Liu, H. W. (2006) Linear aglycones are the substrates for glycosyltransferase DesVII in methymycin biosynthesis: Analysis and implications, *J. Am. Chem. Soc. 128*, 5606–5607.

79. Katz, L., and Donadio, S. (1993) Polyketide synthesis: Prospects for hybrid antibiotics, *Annu. Rev. Microbiol. 47*, 875–912.

80. Gaisser, S., Bohm, G. A., Cortes, J., and Leadlay, P. F. (1997) Analysis of seven genes from the eryAI/-eryK region of the erythromycin biosynthetic gene cluster in *Saccharopolyspora erythraea*, *Mol. Gen. Genet. 256*, 239–251.

81. Summers, R. G., Donadio, S., Staver, M. J., WendtPienkowski, E., Hutchinson, C. R., and Katz, L. (1997) Sequencing and mutagenesis of genes from the erythromycin biosynthetic gene cluster of *Saccharopolyspora erythraea* that are involved in L-mycarose and D-desosamine production, *Microbiology-UK 143*, 3251–3262.

82. Lee, H. Y., Chung, H. S., Hang, C., Khosla, C., Walsh, C. T., Kahne, D., and Walker, S. (2004) Reconstitution and characterization of a new desosaminyl transferase, EryCIII, from the erythromycin biosynthetic pathway, *J. Am. Chem. Soc. 126*, 9924–9925.

83. Yuan, Y., Chung, H. S., Leimkuhler, C., Walsh, C. T., Kahne, D., and Walker, S. (2005) In vitro reconstitution of EryCIII activity for the preparation of unnatural macrolides, *J. Am. Chem. Soc. 127*, 14128–14129.

84. Doumith, M., Legrand, R., Lang, C., Salas, J. A., and Raynal, M. C. (1999) Interspecies complementation in *Saccharopolyspora erythraea*: Elucidation of the function of oleP1, oleG1 and oleG2 from the oleandomycin biosynthetic gene cluster of *Streptomyces antibioticus* and generation of new erythromycin derivatives, *Mol. Microbiol. 34*, 1039–1048.

85. Quiros, L. M., Aguirrezabalaga, I., Olano, C., Mendez, C., and Salas, J. A. (1998) Two glycosyltransferases and a glycosidase are involved in oleandomycin modification during its biosynthesis by *Streptomyces antibioticus*, *Mol. Microbiol.* 28, 1177–1185.

86. Hernandez, C., Olano, C., Mendez, C., and Salas, J. A. (1993) Characterization of a *Streptomyces antibioticus* gene cluster encoding a glycosyltransferase involved in oleandomycin inactivation, *Gene 134*, 139–140.

87. Quiros, L. M., and Salas, J. A. (1995) Biosynthesis of the macrolide oleandomycin by *Streptomyces antibioticus*. Purification and kinetic characterization of an oleandomycin glucosyltransferase, *J. Biol. Chem.* 270, 18234–18239.

88. Sasaki, J., Mizoue, K., Morimoto, S., and Omura, S. (1996) Microbial glycosylation of macrolide antibiotics by *Streptomyces hygroscopicus* ATCC 31080 and distribution of a macrolide glycosyl transferase in several *Streptomyces* strains, *J. Antibiot. (Tokyo) 49*, 1110–1118.

89. Quiros, L. M., Carbajo, R. J., Brana, A. F., and Salas, J. A. (2000) Glycosylation of macrolide antibiotics. Purification and kinetic studies of a macrolide glycosyltransferase from Streptomyces antibioticus, *J. Biol. Chem.* 275, 11713–11720.

90. Yang, M., Brazier, M., Edwards, R., and Davis, B. G. (2005) High-throughput mass-spectrometry monitoring for multisubstrate enzymes: Determining the kinetic parameters and catalytic activities of glycosyltransferases, *Chembiochem 6*, 346–357.

91. Yang, M., Proctor, M. R., Bolam, D. N., Errey, J. C., Field, R. A., Gilbert, H. J., and Davis, B. G. (2005) Probing the breadth of macrolide glycosyltransferases: *In vitro* remodeling of a polyketide antibiotic creates active bacterial uptake and enhances potency, *J. Am. Chem. Soc. 127*, 9336–9337.

92. Oberthür, M., Leimkuhler, C., Kruger, R. G., Lu, W., Walsh, C. T., and Kahne, D. (2005) A systematic investigation of the synthetic utility of glycopeptide glycosyltransferases, *J. Am. Chem. Soc. 127*, 10747–10752.

93. Omura, S., and Crump, A. (2004) The life and times of ivermectin — A success story, *Nat. Rev. Microbiol. 2*, 984–989.

94. Burg, R. W., Miller, B. M., Baker, E. E., Birnbaum, J., Currie, S. A., Hartman, R., Kong, Y. L., Monaghan, R. L., Olson, G., Putter, I., Tunac, J. B., Wallick, H., Stapley, E. O., Oiwa, R., and Omura, S. 1979 Avermectins, new family of potent anthelmintic agents: Producing organism and fermentation, *Antimicrob. Agents Chemother. 15*, 361–367.

95. Ikeda, H., Nonomiya, T., Usami, M., Ohta, T., and Omura, S. (1999) Organization of the biosynthetic gene cluster for the polyketide anthelmintic macrolide avermectin in *Streptomyces avermitilis*, *Proc. Natl. Acad. Sci. USA 96*, 9509–9514.

96. Schulman, M. D., Acton, S. L., Valentino, D. L., and Arison, B. H. (1990) Purification and identification of dTDP-oleandrose, the precursor of the oleandrose units of the avermectins, *J. Biol. Chem. 265*, 16965–16970.

97. Wohlert, S., Lomovskaya, N., Kulowski, K., Fonstein, L., Occi, J. L., Gewain, K. M., MacNeil, D. J., and Hutchinson, C. R. (2001) Insights about the biosynthesis of the avermectin deoxysugar L-oleandrose through heterologous expression of *Streptomyces avermitilis* deoxysugar genes in *Streptomyces lividans*, *Chem. Biol. 8*, 681–700.

98. Luzhetskyy, A., Fedoryshyn, M., Durr, C., Taguchi, T., Novikov, V., and Bechthold, A. (2005) Iteratively acting glycosyltransferases involved in the hexasaccharide biosynthesis of landomycin A, *Chem. Biol. 12*, 725–729.

99. Irschik, H., Jansen, R., Gerth, K., Hofle, G., and Reichenbach, H. (1987) The sorangicins, novel and powerful inhibitors of eubacterial RNA polymerase isolated from myxobacteria, *J. Antibiot. (Tokyo) 40*, 7–13.

100. Campbell, E. A., Pavlova, O., Zenkin, N., Leon, F., Irschik, H., Jansen, R., Severinov, K., and Darst, S. A. (2005) Structural, functional, and genetic analysis of sorangicin inhibition of bacterial RNA polymerase, *EMBO J. 24*, 674–682.

101. Campbell, E. A., Korzheva, N., Mustaev, A., Murakami, K., Nair, S., Goldfarb, A., and Darst, S. A. (2001) Structural mechanism for rifampicin inhibition of bacterial RNA polymerase, *Cell 104*, 901–912.

102. Kopp, M., Rupprath, C., Irschik, H., Bechthold, A., Elling, L., and Muller, R. (2007) SorF: A glycosyltransferase with promiscuous donor substrate specificity in vitro, *Chembiochem 8*, 813–819.

103. Weigel, L. M., Clewell, D. B., Gill, S. R., Clark, N. C., McDougal, L. K., Flannagan, S. E., Kolonay, J. F., Shetty, J., Killgore, G. E., and Tenover, F. C. (2003) Genetic analysis of a high-level vancomycin-resistant isolate of *Staphylococcus aureus*, *Science 302*, 1569–1571.

104. Chang, S., Sievert, D. M., Hageman, J. C., Boulton, M. L., Tenover, F. C., Downes, F. P., Shah, S., Rudrik, J. T., Pupp, G. R., Brown, W. J., Cardo, D., and Fridkin, S. K. (2003) Infection with vancomycin-resistant *Staphylococcus aureus* containing the vanA resistance gene, *N. Engl. J. Med. 348*, 1342–1347.

105. Malabarba, A., and Ciabatti, R. (2001) Glycopeptide derivatives, *Curr. Med. Chem. 8*, 1759–1773.

106. Pace, J. L., and Yang, G. (2006) Glycopeptides: Update on an old successful antibiotic class, *Biochem. Pharmacol. 71*, 968–980.

107. Van Bambeke, F. (2006) Glycopeptides and glycodepsipeptides in clinical development: A comparative review of their antibacterial spectrum, pharmacokinetics and clinical efficacy, *Curr. Opin. Invest. Drugs 7*, 740–749.

108. Losey, H. C., Jiang, J., Biggins, J. B., Oberthur, M., Ye, X. Y., Dong, S. D., Kahne, D., Thorson, J. S., and Walsh, C. T. (2002) Incorporation of glucose analogs by GtfE and GtfD from the vancomycin biosynthetic pathway to generate variant glycopeptides, *Chem. Biol. 9*, 1305–1314.

109. Ge, M., Thompson, C., and Kahne, D. (1998) Reconstruction of vancomycin by chemical glycosylation of the pseudoaglycon, *J. Am. Chem. Soc. 120*, 11014–11015.

110. Thompson, C., Ge, M., and Kahne, D. (1999) Synthesis of vancomycin from the aglycon, *J. Am. Chem. Soc. 121*, 1237–1244.

111. Preobrazhenskaya, M. N., and Olsufyeva, E. N. (2004) Patents on glycopeptides of the vancomycin family and their derivatives as antimicrobials: January 1999–June 2003, *Expert Opin. Ther. Pat. 14*, 141–173.

112. van Wageningen, A. M., Kirkpatrick, P. N., Williams, D. H., Harris, B. R., Kershaw, J. K., Lennard, N. J., Jones, M., Jones, S. J., and Solenberg, P. J. (1998) Sequencing and analysis of genes involved in the biosynthesis of a vancomycin group antibiotic, *Chem. Biol. 5*, 155–162.

113. Losey, H. C., Peczuh, M. W., Chen, Z., Eggert, U. S., Dong, S. D., Pelczer, I., Kahne, D., and Walsh, C. T. (2001) Tandem action of glycosyltransferases in the maturation of vancomycin and teicoplanin aglycones: Novel glycopeptides, *Biochemistry 40*, 4745–4755.

114. Lu, W., Oberthur, M., Leimkuhler, C., Tao, J., Kahne, D., and Walsh, C. T. (2004) Characterization of a regiospecific epivancosaminyl transferase GtfA and enzymatic reconstitution of the antibiotic chloroeremomycin, *Proc. Natl. Acad. Sci. USA 101*, 4390–4395.

115. Freel Meyers, C. L., Oberthur, M., Anderson, J. W., Kahne, D., and Walsh, C. T. (2003) Initial characterization of novobiocic acid noviosyl transferase activity of NovM in biosynthesis of the antibiotic novobiocin, *Biochemistry 42*, 4179–4189.

116. Nagarajan, R. (1993) Structure-activity relationships of vancomycin-type glycopeptide antibiotics, *J. Antibiot. (Tokyo) 46*, 1181–1195.

117. Li, T. L., Huang, F., Haydock, S. F., Mironenko, T., Leadlay, P. F., and Spencer, J. B. (2004) Biosynthetic gene cluster of the glycopeptide antibiotic teicoplanin: Characterization of two glycosyltransferases and the key acyltransferase, *Chem. Biol. 11*, 107–119.

118. Truman, A. W., Robinson, L., and Spencer, J. B. (2006) Identification of a deacetylase involved in the maturation of teicoplanin, *Chembiochem 7*, 1670–1675.

119. Bililign, T., Hyun, C. G., Williams, J. S., Czisny, A. M., and Thorson, J. S. (2004) The hedamycin locus implicates a novel aromatic PKS priming mechanism, *Chem. Biol. 11*, 959–969.

120. Durr, C., Hoffmeister, D., Wohlert, S. E., Ichinose, K., Weber, M., Von Mulert, U., Thorson, J. S., and Bechthold, A. (2004) The glycosyltransferase UrdGT2 catalyzes both C- and O-glycosidic sugar transfers, *Angew. Chem. Int. Ed. Engl. 43*, 2962–2965.

121. Mittler, M., Bechthold, A., and Schulz, G. E. (2007) Structure and action of the C–C bond-forming glycosyltransferase UrdGT2 involved in the biosynthesis of the antibiotic urdamycin, *J. Mol. Biol. 372*, 67–76.

122. Gormley, N. A., Orphanides, G., Meyer, A., Cullis, P. M., and Maxwell, A. (1996) The interaction of coumarin antibiotics with fragments of DNA gyrase B protein, *Biochemistry 35*, 5083–5092.

123. Marcu, M. G., Chadli, A., Bouhouche, I., Catelli, M., and Neckers, L. M. (2000) The heat shock protein 90 antagonist novobiocin interacts with a previously unrecognized ATP-binding domain in the carboxyl terminus of the chaperone, *J. Biol. Chem. 275*, 37181–37186.

124. Zhang, H., and Burrows, F. (2004) Targeting multiple signal transduction pathways through inhibition of Hsp90, *J. Mol. Med. 82*, 488–499.

125. Hanahan, D., and Weinberg, R. A. (2000) The hallmarks of cancer, *Cell 100*, 57–70.

126. Steffensky, M., Muhlenweg, A., Wang, Z. X., Li, S. M., and Heide, L. (2000) Identification of the novobiocin biosynthetic gene cluster of *Streptomyces spheroides* NCIB 11891, *Antimicrob. Agents Chemother. 44*, 1214–1222.

127. Pacholec, M., Freel Meyers, C. L., Oberthur, M., Kahne, D., and Walsh, C. T. (2005) Characterization of the aminocoumarin ligase SimL from the simocyclinone pathway and tandem incubation with NovM,P,N from the novobiocin pathway, *Biochemistry 44*, 4949–4956.

128. Freel Meyers, C. L., Oberthur, M., Heide, L., Kahne, D., and Walsh, C. T. (2004) Assembly of dimeric variants of coumermycins by tandem action of the four biosynthetic enzymes CouL, CouM, CouP, and NovN, *Biochemistry 43*, 15022–15036.

129. Thorson, J. S., Sievers, E. L., Ahlert, J., Shepard, E., Whitwam, R. E., Onwueme, K. C., and Ruppen, M. (2000) Understanding and exploiting nature's chemical arsenal: The past, present and future of calicheamicin research, *Curr. Pharm. Des. 6*, 1841–1879.

130. Devoss, J. J., Hangeland, J. J., and Townsend, C. A. (1990) Characterization of the in vitro cyclization chemistry of calicheamicin and its relation to DNA cleavage, *J. Am. Chem. Soc. 112*, 4554–4556.

131. Watanabe, C. M., Supekova, L., and Schultz, P. G. (2002) Transcriptional effects of the potent enediyne anti-cancer agent Calicheamicin γ_1^I, *Chem. Biol. 9*, 245–251.

132. Sievers, E. L., and Linenberger, M. (2001) Mylotarg: Antibody-targeted chemotherapy comes of age, *Curr. Opin. Oncol. 13*, 522–527.

133. Ahlert, J., Shepard, E., Lomovskaya, N., Zazopoulos, E., Staffa, A., Bachmann, B. O., Huang, K., Fonstein, L., Czisny, A., Whitwam, R. E., Farnet, C. M., and Thorson, J. S. (2002) The calicheamicin gene cluster and its iterative type I enediyne PKS, *Science 297*, 1173–1176.

134. Shindo, K., Kamishohara, M., Odagawa, A., Matsuoka, M., and Kawai, H. (1993) Vicenistatin, a novel 20-membered macrocyclic lactam antitumor antibiotic. *J. Antibiot. (Tokyo) 46*, 1076–1081.

135. Ogasawara, Y., Katayama, K., Minami, A., Otsuka, M., Eguchi, T., and Kakinuma, K. (2004) Cloning, sequencing, and functional analysis of the biosynthetic gene cluster of macrolactam antibiotic vicenistatin in *Streptomyces halstedii*, *Chem. Biol. 11*, 79–86.

136. Minami, A., Kakinuma, K., and Eguchi, T. (2005) Aglycon switch approach toward unnatural glycosides from natural glycoside with glycosyltransferase VinC, *Tetrahedr. Lett. 46*, 6187–6190.

137. Minami, A., Uchida, R., Eguchi, T., and Kakinuma, K. (2005) Enzymatic approach to unnatural glycosides with diverse aglycon scaffolds using glycosyltransferase VinC, *J. Am. Chem. Soc. 127*, 6148–6149.

138. Minami, A., and Eguchi, T. (2007) Substrate flexibility of vicenisaminyltransferase VinC involved in the biosynthesis of vicenistatin, *J. Am. Chem. Soc. 129*, 5102–5107.

139. Hutchinson, C. R. (1997) Biosynthetic studies of daunorubicin and tetracenomycin C, *Chem. Rev. 97*, 2525–2536.

140. Otten, S. L., Stutzman-Engwall, K. J., and Hutchinson, C. R. (1990) Cloning and expression of daunorubicin biosynthesis genes from *Streptomyces peucetius* and *S. peucetius* subsp. *caesius*, *J. Bacteriol. 172*, 3427–3434.

141. Giannini, G. (2002) Fluorinated anthracyclines: Synthesis and biological activity, *Curr. Med. Chem. 9*, 687–712.

142. Larsen, A. K., Escargueil, A. E., and Skladanowski, A. (2003) Catalytic topoisomerase II inhibitors in cancer therapy, *Pharmacol. Ther.* 99, 167–181.

143. Minotti, G., Licata, S., Saponiero, A., Menna, P., Calafiore, A. M., Di Giammarco, G., Liberi, G., Animati, F., Cipollone, A., Manzini, S., and Maggi, C. A. (2000) Anthracycline metabolism and toxicity in human myocardium: Comparisons between doxorubicin, epirubicin, and a novel disaccharide analogue with a reduced level of formation and [4Fe-4S] reactivity of its secondary alcohol metabolite, *Chem. Res. Toxicol.* 13, 1336–1341.

144. Temperini, C., Cirilli, M., Aschi, M., and Ughetto, G. (2005) Role of the amino sugar in the DNA binding of disaccharide anthracyclines: crystal structure of the complex MAR70/d(CGATCG), *Bioorg. Med. Chem.* 13, 1673–1679.

145. Animati, F., Arcamone, F., Bigioni, M., Capranico, G., Caserini, C., De Cesare, M., Lombardi, P., Pratesi, G., Salvatore, C., Supino, R., and Zunino, F. (1996) Biochemical and pharmacological activity of novel 8-fluoroanthracyclines: Influence of stereochemistry and conformation, *Mol. Pharmacol.* 50, 603–609.

146. Gate, L., Couvreur, P., Nguyen-Ba, G., and Tapiero, H. (2003) N-methylation of anthracyclines modulates their cytotoxicity and pharmacokinetic in wild type and multidrug resistant cells, *Biomed. Pharmacother.* 57, 301–308.

147. Licata, S., Saponiero, A., Mordente, A., and Minotti, G. (2000) Doxorubicin metabolism and toxicity in human myocardium: Role of cytoplasmic deglycosidation and carbonyl reduction, *Chem. Res. Toxicol.* 13, 414–420.

148. Schimmel, K. J., Richel, D. J., van den Brink, R. B., and Guchelaar, H. J. (2004) Cardiotoxicity of cytotoxic drugs, *Cancer Treat. Rev.* 30, 181–191.

149. Qu, X., Wan, C., Becker, H. C., Zhong, D., and Zewail, A. H. (2001) The anticancer drug-DNA complex: Femtosecond primary dynamics for anthracycline antibiotics function, *Proc. Natl. Acad. Sci. USA* 98, 14212–14217.

150. Zhong, D., Pal, S. K., Wan, C., and Zewail, A. H. (2001) Femtosecond dynamics of a drug-protein complex: Daunomycin with Apo riboflavin-binding protein, *Proc. Natl. Acad. Sci. USA* 98, 11873–11878.

151. Raty, K., Kantola, J., Hautala, A., Hakala, J., Ylihonko, K., and Mantsala, P. (2002) Cloning and characterization of *Streptomyces galilaeus* aclacinomycins polyketide synthase (PKS) cluster, *Gene* 293, 115–122.

152. Lu, W., Leimkuhler, C., Oberthur, M., Kahne, D., and Walsh, C. T. (2004) AknK is an L-2-deoxyfucosyltransferase in the biosynthesis of the anthracycline aclacinomycin A, *Biochemistry* 43, 4548–4558.

153. Lu, W., Leimkuhler, C., Gatto, G. J., Jr., Kruger, R. G., Oberthur, M., Kahne, D., and Walsh, C. T. (2005) AknT is an activating protein for the glycosyltransferase AknS in L-aminodeoxysugar transfer to the aglycone of aclacinomycin A, *Chem. Biol.* 12, 527–534.

154. Leimkuhler, C., Fridman, M., Lupoli, T., Walker, S., Walsh, C. T., and Kahne, D. (2007) Characterization of rhodosaminyl transfer by the AknS/AknT glycosylation complex and its use in reconstituting the biosynthetic pathway of aclacinomycin A, *J. Am. Chem. Soc.* 129, 10546–10550.

155.   Sianidis, G., Wohlert, S. E., Pozidis, C., Karamanou, S., Luzhetskyy, A., Vente, A., and
       Economou, A. (2006) Cloning, purification and characterization of a functional anthra-
       cycline glycosyltransferase, *J. Biotechnol. 125*, 425–433.
156.   Faust, B., Hoffmeister, D., Weitnauer, G., Westrich, L., Haag, S., Schneider, P., Decker,
       H., Kunzel, E., Rohr, J., and Bechthold, A. (2000) Two new tailoring enzymes, a
       glycosyltransferase and an oxygenase, involved in biosynthesis of the angucycline
       antibiotic urdamycin A in *Streptomyces fradiae* Tu2717, *Microbiology 146*Pt.1,
       147–154.
157.   Trefzer, A., Hoffmeister, D., Kunzel, E., Stockert, S., Weitnauer, G., Westrich, L., Rix, U.,
       Fuchser, J., Bindseil, K. U., Rohr, J., and Bechthold, A. (2000) Function of glycosyl-
       transferase genes involved in urdamycin A biosynthesis, *Chem. Biol. 7*, 133–142.
158.   Hoffmeister, D., Ichinose, K., and Bechthold, A. (2001) Two sequence elements of
       glycosyltransferases involved in urdamycin biosynthesis are responsible for substrate
       specificity and enzymatic activity, *Chem. Biol. 8*, 557–567.
159.   Hoffmeister, D., Wilkinson, B., Foster, G., Sidebottom, P. J., Ichinose, K., and Bechthold,
       A. (2002) Engineered urdamycin glycosyltransferases are broadened and altered in
       substrate specificity, *Chem. Biol. 9*, 287–295.
160.   Kubo, A., Arai, Y., Nagashima, S., and Yoshikawa, T. (2004) Alteration of sugar donor
       specificities of plant glycosyltransferases by a single point mutation, *Arch. Biochem.
       Biophys. 429*, 198–203.
161.   Kohara, A., Nakajima, C., Yoshida, S., and Muranaka, T. (2007) Characterization and
       engineering of glycosyltransferases responsible for steroid saponin biosynthesis in
       Solanaceous plants, *Phytochemistry 68*, 478–486.
162.   Hancock, S. M., Vaughan, M. D., and Withers, S. G. (2006) Engineering of glycosidases
       and glycosyltransferases, *Curr. Opin. Chem. Biol. 10*, 509–519.
163.   Qasba, P. K., Ramakrishnan, B., and Boeggeman, E. (2006) Mutant glycosyltransferases
       assist in the development of a targeted drug delivery system and contrast agents for MRI,
       AAPS J. 8, E190–E195.
164.   Ramakrishnan, B., Boeggeman, E., and Qasba, P. K. (2005) Mutation of arginine 228 to
       lysine enhances the glucosyltransferase activity of bovine β-1,4-galactosyltransferase I,
       Biochemistry 44, 3202–3210.
165.   Ramakrishnan, B., and Qasba, P. K. (2002) Structure-based design of β1,4-galactosyl-
       transferase I (β4Gal-T1) with equally efficient *N*-acetylgalactosaminyltransferase activity:
       Point mutation broadens β4Gal-T1 donor specificity, *J. Biol. Chem. 277*, 20833–20839.
166.   Khidekel, N., Arndt, S., Lamarre-Vincent, N., Lippert, A., Poulin-Kerstien, K. G.,
       Ramakrishnan, B., Qasba, P. K., and Hsieh-Wilson, L. C. (2003) A chemoenzymatic
       approach toward the rapid and sensitive detection of *O*-GlcNAc posttranslational
       modifications, *J. Am. Chem. Soc. 125*, 16162–16163.
167.   He, X. Z., Wang, X., and Dixon, R. A. (2006) Mutational analysis of the *Medicago
       glycosyltransferase* UGT71G1 reveals residues that control regioselectivity for (iso)
       flavonoid glycosylation, *J. Biol. Chem. 281*, 34441–34447.
168.   Aharoni, A., Thieme, K., Chiu, C. P., Buchini, S., Lairson, L. L., Chen, H., Strynadka,
       N. C., Wakarchuk, W. W., and Withers, S. G. (2006) High-throughput screening

methodology for the directed evolution of glycosyltransferases, *Nat. Methods 3*, 609–614.

169. Williams, G. J., Zhang, C., and Thorson, J. S. (2007) Directed evolution of a natural product glycosyltransferase, *Nat. Chem. Biol. 3*, 657–662.

170. Collier, A. C., Tingle, M. D., Keelan, J. A., Paxton, J. W., and Mitchell, M. D. (2000) A highly sensitive fluorescent microplate method for the determination of UDP-glucuronosyl transferase activity in tissues and placental cell lines, *Drug Metab. Dispos. 28*, 1184–1186.

171. Mayer, C., Jakeman, D. L., Mah, M., Karjala, G., Gal, L., Warren, R. A., and Withers, S. G. (2001) Directed evolution of new glycosynthases from *Agrobacterium* beta-glucosidase: A general screen to detect enzymes for oligosaccharide synthesis, *Chem. Biol. 8*, 437–443.

COMBINATORIAL AND EVOLUTIONARY DESIGN OF BIOSYNTHETIC REACTION SEQUENCES

By ETHAN T. JOHNSON, ERIK HOLTZAPPLE, and CLAUDIA SCHMIDT-DANNERT, *Department of Biochemistry, Molecular Biology and Biophysics, University of Minnesota, St. Paul, Minnesota 55108*

CONTENTS

_____

Advances in Enzymology and Related Areas of Molecular Biology, Volume 76
Edited by Eric J. Toone   Copyright © 2009 by John Wiley & Sons, Inc.

I. Introduction

Recent advances in protein engineering and microbial genomics have created new opportunities for the biological synthesis of medical and industrial compounds. As petroleum-based synthesis becomes more expensive because of environmental concerns and increases in the price of oil, chemical companies have become interested in the use of engineered microbial systems for the production of fine chemicals. Assembly and manipulation of heterologous biosynthetic reaction sequences in a host organism allow for an efficient and cost-effective multistep enzymatic synthesis in a single bioreactor for the production of complex compounds (1–4).

Traditionally, microorganisms isolated from the environment have been used for the production of industrial and pharmaceutical chemicals. Classic strain improvement strategies using iterative rounds of random mutagenesis and selection have been applied to develop strains capable of overproducing large amounts of a desired metabolite. This approach targets the whole organism and often is not concerned with the molecular reactions that drive this process. However, with the development of recombinant DNA technologies and increasing knowledge about the underlying molecular principles of metabolic processes, it has become possible to engineer desirable metabolic traits into microbial production hosts using rational design strategies (5, 6). This new field of metabolic engineering originated from the heterologous expression of single enzymes in a production host and has developed into the expression of multienzyme pathways composed of enzymes from several organisms. The ability to combine enzymatic reactions into new biosynthetic reaction sequences makes it possible to use microbial cells as chemical factories for the production of complex natural and unnatural small molecules. The wealth of available genomic sequences and new metagenomic approaches provide an enormous resource for the identification of enzyme functions that can be incorporated into biosynthetic reaction sequences for the production of novel small-molecule products or to diversify existing chemical scaffolds.

In this chapter, we provide an introduction to the tools used in assembling biosynthetic reactions in a microbial host by discussing synthesis of several types of natural products in engineered microbial systems. We discuss both rational and evolutionary strategies for pathway design and optimization, and specific emphasis is placed on problems encountered with current strategies that limit production and diversity of produced small molecules. Discussions will follow the general strategy of pathway design and optimization:

1. Identification of metabolic enzymes for pathway assembly from known pathways in genomic databases or by characterization of previously unknown enzyme functions.
2. Following the initial reaction sequence design, biosynthetic enzymes are expressed in a heterologous microbial host, and activity and production levels are established.
3. Finally, the pathway is optimized and integrated into the host's metabolic network for optimal metabolic flux and maximum production.

II. Genomic and Metagenomic Tools

A. SEQUENCE DATABASES

Bioinformatics has become a valuable tool in obtaining biological information for protein engineering and pathway design. Genomics, defined as the study of all the genetic material of an organism, is an amalgamation of molecular and computational biology and provides a structure for understanding the genetic information of an organism. At the time of writing this chapter, 534 completed genome projects have been published, 1095 microbial genome sequencing projects that are ongoing and 73 metagenome projects (both ongoing and completed) according to the GOLD database (genomesonline.org). Genome sequences and annotations can give an immediate view into the metabolic pathways of an organism. Metagenome projects have yielded a glimpse into unknown biochemical pathways of unculturable microorganisms, indicating that we have so far barely scratched the surface of the biosynthetic potential found in nature (7).

One of the most useful bioinformatics tools for gene identification is the NCBI's (National Center for Biotechnology Information) BLAST (Basic Local Alignment Search Tool), an algorithm for comparing DNA or peptide sequence(s) (8). Peptide or DNA sequences are compared to characterized enzymes or putatively assigned open reading frames (ORFs) from annotated genomes; however, caution is necessary when drawing conclusions from putatively assigned ORFs as "genome rot" can occur if assignments of gene function are based solely on other putatively assigned genes. Gene assignments in genomes are scored based on alignments to annotated ORFs in GenBank (sequence database at NCBI composed of experimentally characterized genes as well as genomes). Several variations of BLAST are available that compare DNA or protein sequences to the database. One of these, tblastn, queries a peptide sequence against all six possible reading frames of the target DNA segment and is ideal for finding novel proteins from unannotated or unfinished genome sequences.

The location of the gene on the chromosome in relation to other genes that may be involved in a biosynthetic pathway may also aid in the search for biosynthetic enzymes as functional linkage often can be inferred if the neighboring genes are found in several genomes. Gene clusters (also called operons) are collections of genes in close proximity that are associated with a biological process. The Institute for Genomic Research (TIGR, now called JCVI) has a tool on TIGR's Comprehensive Microbial Resource (CMR) called Regional Display Viewer that shows a view of an ORF in relation to other genes surrounding it (9). As an example of the utility of this tool, the regional view of the genes on a chromosome aided in finding biosynthetic enzymes involved in the biosynthesis of acyclic xanthophylls (carotenoids with oxygen-containing functional groups) in *Staphylococcus aureus* (see Figure 1). A new desaturase homolog (*crtOx*) was identified in *Staphylococcus* located near *crtM* and *crtN*, two well-known carotenoid enzymes. The novel CrtOx was able to add oxygen-containing functional groups to carotenoid molecules (10).

Conversely, knowledge of established catalytic reactions in experimentally described microorganisms can indicate which ORFs may have the capability to synthesize the desired product if homologous genes are found in the genome of interest. Prior knowledge of a natural product biosynthesis can provide insight in discovering conserved enzymes. Here, we present two examples that use microbial genome sequences to identify novel metabolic pathways.

Marinobacter hydrocarbonoclasticus makes isoprenoid wax ester compounds when grown with phytol as the sole carbon source under nitrogen and/ or phosphorus limiting growth conditions (11). Wax esters are neutral lipids composed of fatty alcohols and fatty acids. At the time of the first description of isoprenoid esters, no enzymes were known to catalyze the formation of wax esters in prokaryotes. A new type of acyltransferase, a bifunctional wax ester synthase/diacyl glycerol acyltransferase (WS/DGAT) catalyzing synthesis of acyl wax esters from acyl-CoA and fatty alcohols, was characterized from the prokaryote *Acinetobacter bayli* ADP1 (12). Based on the

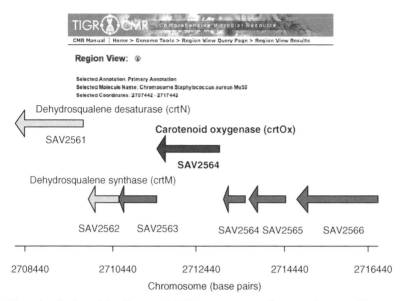

Figure 1. Region of the *S. aureus* Mu50 genome surrounding several carotenoid genes illustrates the regional display tool from the TIGR CMR Web tool for observing neighboring ORFs. Arrows correspond to ORFs labeled with their gene identifications. The symbols *crtM* and *crtN* (gold arrows) are previously characterized carotenoid genes from *S. aureus* known to make diapolycopene (C30) carotenoids. A role for *crtOx* (violet arrow) in the carotenoid pathway was suggested by its proximity to *crtM* and *crtN* and subsequent biochemical characterization has identified CrtOx as a carotenoid oxygenase. Putative or hypothetical ORFs (gray arrows) are also shown. (See insert for color representation.)

identification of this enzyme function, it was hypothesized that isoprenoid wax ester formation in *Marinobacter* would involve a similar mechanism. The peptide sequence of the *Acinetobacter* WS/DGAT was used to search the draft genome sequence of *Marinobacter aquaeolei*, an organism closely related to the isoprenoid wax ester producing *M. hydrocarbonoclasticus*, for putative WS sequences (13). Four putative ORFs with similarity to the *Acinetobacter* enzyme were identified. With this knowledge, two WS sequences capable of producing an isoprenoid wax from phytanoyl-CoA and phytol were amplified by polymerase chain reaction (PCR) from *M. hydrocarbonoclasticus* (14).

Genomics methods also may help in elucidating the biosynthesis of a natural product even in cases when there is no previous knowledge about the specific biosynthetic pathway. Identification of the biosynthesis of the anticancer peptide drugs patellamide A and C by *Prochloron didemni*, a cyanobacterial endosymbiont of ascidians (sea squirts), is an example of how genome information was crucial in determining the biosynthesis of a natural product. Initially, researchers thought patellamide was a nonribosomal peptide synthase (NRPS)–derived compound; however, DNA probes of conserved NRPS domains failed to show any evidence for NRPS modules normally associated with peptide natural products. Surprisingly, the peptide sequences for patellamide C and A were found directly encoded in the genome when the draft genome sequence of *P. didemni* was completed. A protease adjacent to the peptide sequences was confirmed to cleave the patellamide peptide fragments from a leader sequence. To support the researchers' claim, patellamide A and C can be produced in recombinant *Escherichia coli* by expression of the patellamide biosynthesis gene cluster (15).

B.   METABOLIC PATHWAY DATABASES

A number of specialized databases are available for searching genome sequences for metabolic pathway information. Several very useful database collections for pathway design and discovery are the KEGG databases, the MetaCyc, BioCyc, and EcoCyc databases by the Stanford Research Institute (SRI International), the University of Minnesota Biocatalysis/Biodegradation database, and a metagenomics database under development named CAMERA.

1.   KEGG: Kyoto Encyclopedia of Genes and Genomes

KEGG is a collection of four databases (KEGG PATHWAY, KEGG GENES, KEGG LIGAND, and KEGG BRITE) linking cellular processes together to gain a better understanding of biological systems (16). It contains information on over 443 bacteria, 37 archaea, and 80 eukaryotes. The four database networks are crosslinked to query several fields for comprehensive knowledge retrieval on a particular organism or a biochemical reaction or metabolite.

KEGG PATHWAY contains maps and displays information of molecular and biochemical reaction networks, including metabolic pathway maps. A collection of gene catalogs from publicly available sources for complete high-quality genomes (GENES gene catalog), draft genomes (DGENES), expressed sequence tag (EST) contigs (EGENES), viral genomes

(VGENES), and organelle genomes (OGENES) can be queried in KEGG GENES. The KEGG LIGAND database contains links to enzyme nomenclature as well as to chemical structures of known metabolic compounds, pharmaceutical, and environmental compounds. All chemical structures are entered manually, computationally verified, and continuously updated. Currently, the database contains 14,549 entries. The newest addition to the KEGG suite of databases is KEGG BRITE. This database provides hierarchical classification of functions in biological systems. It includes information on genes and proteins, chemical compounds and reactions, protein ligands, drugs and drug interactions, diseases, cells, and organisms.

2. BioCyc with MetaCyc and EcoCyc

BioCyc is a collection of 260 genome and pathway databases (17). It includes three intensively curated databases: BioCyc, EcoCyc, and Meta-Cyc. Biocyc is a database that displays all components of individual metabolic pathways, including enzymes, intermediate compounds, and cofactors. EcoCyc is a subset of this information that focuses on the reactions of *E. coli* (18). This database is useful for the metabolic engineer using *E. coli* as a host because it provides information specific to *E. coli's* genome, metabolic pathways, transporters, and regulatory networks.

By contrast, MetaCyc is a database that contains information on over 900 experimentally established metabolic pathways of primary and secondary metabolism from over 900 organisms. This database provides information on metabolic pathways, enzymatic reactions, enzymes, chemical compounds, and genes (19). Enzyme information includes substrate specificity, kinetic properties, activators, inhibitors, cofactor requirements, and links to sequence and structure databases. Data are routinely curated from the primary literature by analysts with expertise in biochemistry and molecular biology. This database is a very useful tool for metabolic pathway engineering as it provides reference data sets for predicting putative metabolic pathways of organisms from sequenced genomes, allows comparison of biochemical networks, and serves as a comprehensive encyclopedia of metabolism.

3. UM-BBD-U of MN Biocatalysis/Biodegradation Database

The University of Minnesota's Microbial Biocatalytic Reactions and Biodegradation Pathway Database (UM-BBD) lists information on over 900 compounds, over 600 enzymes, and approximately 1000 chemical reactions for 350 microbial entries (20). The Pathway Prediction System

(PPS) predicts microbial catabolism of organic substrates and other anthropogenic compounds based on established chemical reactions of biochemistry and organic chemistry (20). The chemical structures of all possible outcomes of a catalytic reaction are listed by a score of how likely the reaction would occur.

4. CAMERA (Cyberinfrastructure for Advanced Marine Microbial Ecology Research and Analysis)

A new metagenome database still under development by the J. Craig Venter Institute contains assembled contiguous (contigs) DNA sequence data from ecological samples of mixed microbial communities (7). BLAST can be used to search protein or DNA sequences against data sets of DNA contigs of uncultured microbial communities from ocean samples taken from several locations. Metagenomics has the potential to extrapolate and infer novel biochemical pathways of these previously uncharacterized, unknown microbial species.

III.   Biosynthetic Pathway Design and Diversification

To illustrate the breadth of the problems encountered in engineering pathways in heterologous organisms that must be overcome, we discuss several areas of active research. The first section develops the ideas of combinatorial pathway design using the type I polyketide synthases (PKSs) as an example. Type I PKSs are a focus of engineering efforts because of their linear and modular organization that makes them amenable for many catalytic permutations. Further, recent structural studies of these PKSs are described to show how mechanistic insights guide pathway engineering. In the second section, examples from terpene and carotenoid biosynthesis illustrate evolutionary methods of protein and pathway design for the production of natural and novel compounds.

While we discuss only a few types of molecules, it is important to realize that many efforts are being made in combinatorial and evolutionary pathway engineering. Improvements in pathway design and development of antibiotics are not limited to polyketides (21); bioactive nonribsomal peptides have been produced in recombinant *E. coli* expressing nonribosomal peptide synthases (NRPSs) (22, 23), and alkaloids, another important class of biologically active compounds, are also under intense investigation (24). Structural studies have given insight into the biochemistry of the NRP pathway (25) and hybrid enzymes composed of PKS and NRPS modules have been constructed (26). Additionally, flavonoid pathways have been reconstructed (27–30) and a heterologous porphyrin pathway overexpressed in *E. coli* (31, 32).

A. COMBINATORIAL METHODS: EXAMPLE POLYKETIDE BIOSYNTHESIS

The PKS biosynthetic gene clusters are some of the most explored pathways in combinatorial pathway engineering (for extensive reviews see refs. 33–36). Polyketides are bioactive secondary metabolites produced primarily by filamentous bacteria (actinomycetes) and often are antibacterial compounds. Formed by the successive condensation of small organic acids, such as acetic and malonic acids, the polyketides are a diverse family of compounds whose structures depend on the types and sequences of the enzymatic reactions encoded by the assembly-line PKS enzymes. Generally PKSs are composed of several modules each containing, at the minimum, three functional domains: an acyltransferase (AT), an acyl carrier protein (ACP), and a ketosynthase (KS) (see Figure 2). The AT domain binds an extender unit and transfers it to the ACP, and the KS domain catalyzes the condensation of the growing polyketide chain with the extender unit. Within each module, various combinations of ketoreductase, enoyl reductase, and dehydratase domains reduce the ketide before the chain is extended further. Finally, the polyketide chain is terminated by a thioesterase domain that may also catalyze formation of a macrolactone ring. Additional tailoring enzymes such as oxygenases, halogenases, and glycosyltransferases that are not part of the PKS modules further modify the polyketide product either during or after synthesis (37, 38).

The diversity of the structures stems from the length of the carbon backbone formed by the various types of extender units and polyketide chain modifications. Currently, six types of extender units are known: malonyl-CoA, methylmalonyl-CoA, ethylmalonyl-CoA, methoxymalonyl-ACP, hydroxymalonyl-ACP, and aminomalonyl-ACP (39). It is likely that additional extender units will be discovered as new gene clusters for PKS are characterized biochemically. Currently, there are hundreds of known biosynthetic clusters for PKSs, and this number increases as genomes are sequenced and strategies for detection of bacterial gene clusters encoding PKSs and NRPSs are improved (40). Further, the modular nature of the PKSs has piqued interest in engineering novel combinations of modules to generate diverse compounds.

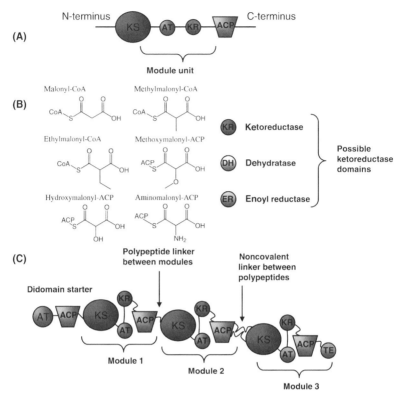

Figure 2.   Type I polyketide synthases as an example of combinatorial biosynthesis. (A) The primary sequence of the multidomain enzyme complex contains several modules each possessing ketosynthase (KS), acyltransferase (AT), and ketoreductase (KR) domains and the initial and final modules containing starter and thioesterase (TE) domains. (B) Several types of extender units and KR domain catalytic activity define the reaction for each module unit. Engineering within each module has led to novel products and reaction sequences. (C) Several polypeptides interact to form the tertiary and quaternary structures that determine the complete reaction sequence. The linker regions that are defined by protein–protein interactions between consecutive modules are key structural elements that may be engineered to reorder reaction sequences for novel polyketides. (See insert for color representation.)

1.   Polyketide Biosynthesis in E. coli

Polyketide biosynthetic pathways can be overexpressed either in the native organism or in a heterologous host. While the titers of natural compounds may be higher in the native organism, the genetic tractability of *E. coli* or yeast is often advantageous for the production of novel and unnatural compounds. Additionally, the native organism likely will not have ideal growth characteristics and unique fermentation conditions will be needed for each type of microbial system. To obtain a robust fermentation condition that is general for many types of compounds and easy to scale up to industrial production, the development of a common heterologous host offers a distinct advantage. Pfeifer et al. (41) engineered *E. coli* to produce the complex polyketide 6-deoxyerythronolide B (6-DEB) by overcoming problems typical of pathway engineering. First, they were able to achieve soluble and functional expression of PKS megasynthases (2 MDa) by lowering the temperature of the induction conditions and achieved posttranslational modification of the ACP domain by coexpression of a phosphopantetheinyl transferase. Second, expression of a propionyl-CoA ligase and propionyl-CoA carboxylase was required to obtain the propionyl-CoA starter unit and the malonyl-CoA extender units not produced by *E. coli* (41, 42). Methyl-malonyl-CoA extender units now also have been expressed in pseudomonads and yeast, increasing the choice of heterologous expression hosts (43, 44).

In some cases, production levels may be increased by changing and improving the heterologous production host or by optimization of the codon frequency and promoter strength. Redesigning the 6-DEB pathway using a codon-optimized version of the synthase for *E. coli* produced the compound efficiently (45). Modification of the fermentation conditions and expression of a thioesterase from yeast have increased the production levels 100-fold (42). Lee et al. (46) have developed a method for high-throughput strain improvement for the production of macrolides by *E. coli*. While yeast and *E. coli* are the most common heterologous production hosts, recombinant biosynthesis of some polyketide compounds may benefit from development of alternative hosts with improved metabolic properties such as *Pseudomonas putida* (47–49).

Production of fully decorated polyketides in heterologous hosts poses another pathway engineering challenge while at the same time offering the opportunity of producing compounds with novel and unnatural scaffold modificiations. Glycosylation of 6-DEB generates the bioactive compound, and for production of the fully decorated compound in *E. coli* the engineered pathway included sugar cassettes based upon the megalomicin gene cluster to generate the specific deoxysugars, a glycosyltransferase to transfer this moiety to the completed polyketide and a recombinant ribonucleic acid (RNA) methyltransferase to confer resistance to the host cell from the mature antibiotic (50). Production of glycosylated natural products in *E. coli* therefore requires the installment of sugar pathways and overexpression of cognate glycosyltransferases able to transfer the specific sugars to the antibiotic molecule (50). Lombo et al. (51) have constructed sugar biosynthetic cassettes to produce branched deoxysugars and expressed elloramycin glycosyltransferase, an enzyme able to transfer sugars to many types of substrates, to generate two tetracenomycin derivatives. Combinations of novel deoxysugars with aglycones (52) and extension to the antibiotic staurosporine (24) provide further examples of the general use of the sugar cassette approach (51).

2. Rearranging and Modification of PKS Assembly Line

The type I PKSs are composed of consecutive modules that determine the biosynthetic reaction sequence (see Figure 2C). This observation has led to the idea that new antibacterial compounds can be constructed by reordering the modules in a combinatorial fashion. The modular organization of the PKSs has invited many attempts at engineering the production of novel polyketides by modification of the biosynthetic pathways by domain inactivation, deletion, or substitution (53–55). The many achievements include substitution of the starter and extender units, alteration of the β-carbon processing (ketoreduction, dehydration, and enoyl reduction), and chain elongation. In one of the first experiments to examine a library of sequences encoding combinations of several modules, more than 50 novel macrolide compounds were generated by combinatorially exchanging the ketoreductase and AT domains of 6-DEB synthase modules with those of the PKS for rapamycin (56). These experiments demonstrated that the major components could be substituted and continue to produce active PKSs. More recently, a general method for the de novo biosynthesis of polyketides was devised by introducing unique restriction sites surrounding highly conserved regions in the AT, KS, and reductase domains (57). The fragments from 14 modules from 8 PKS clusters were rearranged to form 154 unique PKSs and nearly half produced triketide lactones when expressed in E. coli. In a complementary approach, DNA shuffling has been used to construct a library of hybrid PKSs and hints at the construction of large libraries of arbitrarily ordered natural PKS modules (58).

While these experiments reveal the promise of the modular engineering of PKSs, most of the combinations did not produce measurable amounts of polyketides. The problems encountered are typical of pathway engineering studies and the details of the PKSs illustrate several of the principles for pathway construction. Several recent discussions have highlighted the issues that may be limiting the production of these systems, and many of the problems are structural, affecting both the properties of individual domains and the interactions between adjacent catalytic domains (21, 35, 54, 59, 60).

The first requirement for building novel modular PKSs is the engineering of broad substrate recognition into the catalytic active sites of consecutive domains. Barriers to effective processing may be due to poor KS-acylation or the subsequent decarboxylation reaction of carbon–carbon bond formation (61, 62). As the order of modules changes, new structures confront the enzyme modules, and for efficient catalysis to occur the active sites of each domain need to be plastic. The intrinsic substrate specificity needs to be overcome by protein engineering of KS domain active sites or matching downstream acceptors with upstream donors.

3. Protein–Protein Interactions in PKS Biosynthesis

Protein–protein interactions between domains in a module and domains of adjacent modules ensure efficient chain elongation and modification during polyketide biosynthesis. In large part these interactions are determined by the linker regions that connect domains within a single module and the linker region that stabilizes the connection between adjacent modules (see Figure 2C). Immediately downstream of the AT domain is a proline-rich region that appears to wrap around both the AT and KS domains providing a framework for the module, and the KS and AT domains are linked by a three-stranded β-sheet. Proper interactions within the linker region may allow the ACP and the next KS domain to communicate effectively. The ACP domain appears to move extensively to present the growing polyketide chain with the AT and KS domains that have active sites separated by 80 Å. The ACP structure (63, 64) also shows that in order to interact with all the components and to move the large distances the ACP must be compatible with the AT and KS domains. A strategy that crosslinks the ACP to the KS domain may provide further insights into the role of primary amino acid sequence in determining protein interactions between these interfaces (65, 66).

Coordination of modules on two separate polypeptide chains in type I PKSs is mediated by protein–protein interactions of the docking domains on the C-terminus of one module and the N-terminus of the second module. The docking domains individually fold into two helices and together fold to form a tightly coupled four-helix bundle. The intermodular linker regions have a dramatic effect on the activity of the chimeric PKS and show that these regions require nativelike interfaces (67). To study the interaction by combining native and chimeric combinations, Chandran et al. (68) have compared the intermodular interactions using pairs of upstream ACP and downstream KS domains. The study indicates that the domains that natively interact in natural systems are the most active, lending support to the idea that these interfaces are determined by the ACP and KS domains. More recent studies have indicated that the protein–protein interactions that lead to efficient transfer of the polyketide between modules are specific for several key residues in the linker regions (69). These residues are expected to interact with residues in the adjacent module to form the four-helix bundle.

The ramifications of the structural basis for the interaction between consecutive polypeptide modules are enticing for combinatorial polyketide biosynthesis. The specific linkage between two modules would provide the opportunity to design PKS assembly lines where the combined modules are engineered to contain residues that form the docking interactions for efficient substrate channeling between modules.

B.   EVOLUTIONARY METHODS: EXAMPLE ISOPRENOID BIOSYNTHESIS

Production of isoprenoid compounds in metabolically engineered hosts illustrates the use of gene combinations and directed evolution for high-level production of novel and natural compounds. Isoprenoids are a remarkably diverse class of compounds consisting of more than 50,000 structures built from five-carbon building blocks, isopentenyl diphosphate (IPP), and dimethylallyl diphosphate (DMAPP). Many essential oils are combinations of isoprene units, including myrcene ($C_{10}H_{16}$) from bay leaves, limonene ($C_{10}H_{15}$) from lemon oil, and zingiberene ($C_{15}H_{24}$) from ginger. In addition to their role as fragrant compounds, the terpenes, composed of two (C10, monoterpenes), three (C15, sequiterpenes), or four (C20, diterpenes) isoprene units, often have antimicrobial and pharmaceutical properties. The sesquiterpenoid artemesinin is recognized for antimalarial properties and the diterpenoid taxol disrupts mictotubule formation and functions as a chemotherapy agent. Other important isoprenoid classes are those composed of six (C30, triterpenes) and eight (C40, tetraterpenes) isoprene units. Ergosterol is an example of a triterpenoid while most carotenoids are tetraterpenes. Isoprene side chains are found in many metabolites, such as coenzymes Q6–Q10, vitamin E, and chlorophylls.

Isoprenoid pathways have been engineered in *E. coli* and *Saccharomyces cerevisiae. Escherichia coli* synthesizes isoprene units from glyceraldehyde

3-phospate and pyruvate via the deoxyxylulose 5-phosphate (DXP) pathway, while *Saccharomyces* derives isoprene units from acetyl-CoA via the mevalonate pathway (1). Both hosts can condense three isoprene units to farnesyl diphosphate (C15) as a precursor for (ubi)quinones and ergosterol (*S. cerevisiae* only). Engineered biosynthesis of additional isoprenoid compounds involves extension of the native isoprenoid precursor pathway in *E. coli* or *S. cerevisiae* with additional gene functions and, for increased production levels, optimization of flux through the isoprenoid precursor pathway.

Increases in isoprenoid production were obtained by balancing glyceraldehyde 3-phospate and pyruvate levels in *E. coli* (70, 71). In recent work, the native chromosomal promoters for the DXP pathway were replaced with strong promoters increasing the flux through the pathway (72). Because *E. coli's* DXP pathway is under strict metabolic control and overexpression of some of its enzyme functions is not well tolerated by *E. coli* (73), the yeast mevalonate pathway was engineered into *E. coli* to bypass metabolic regulation in this host and improve terpenoid production levels (74). Recent analysis and balancing of the pathway intermediates have increased further terpenoid production in *E. coli* (75).

Isoprenoid production has also been obtained in engineered yeast strains (76, 77). Yeast hosts have the advantage that the P450 enzymes that modify the terpenoid scaffold may be expressed more easily in this host. Ultimately, the choice of a host strain depends on both the precursor supply and the activity of downstream modifying enzymes.

1. Carotenoids

In vitro evolution of key enzyme functions coupled with gene combinations from heterologous organisms has generated biosynthetic pathways for the production of diverse carotenoid molecules (4). Natural carotenoids vary in the length of the carbon backbone, extent of conjugation, and modification of both ends of the molecule, and these properties determine their roles in biology as components of light-harvesting photosynthetic proteins as colorants and antioxidants. Carotenoids are derived by the head-to-head condensation, catalyzed by a carotenoid synthase, of either two C15 farnesyl diphosphate (FPP) molecules or two C20 geranylgeranyl diphosphate (GGDP) molecules to generate C30 or C40 carotenoids, respectively. Carotenoid desaturases then introduce double bonds along the carotenoid backbone to generate a chromophore that absorbs in the visible range, and subsequent cyclization or oxidation reactions of the ends of the linear molecule create a diverse range of additional linear and cyclic carotenoid structures. Directed evolution has been used to create carotenoid molecules with a fully conjugated electron system. Here, a library of carotenoid desaturase variants generated from two shuffled desaturase genes from the genus *Erwinia* was screened for mutants able to produce carotenoids absorbing at longer wavelengths (78, 79). One mutant desaturase was found to introduce six rather than four double bonds into the C40 carotenoid backbone phytoene to produce the fully conjugated carotenoid tetradehydrolycopene. This evolved pathway then was extended with a library of cyclase mutants and screened for variants capable of cyclizing the end groups of carotenoids with extended chromophores. A mutant cyclase was identified in the library that produced a new carotenoid compound, torulene.

Extension of these two evolved pathways with known carotenoid end group modifying enzymes such as carotenoid monooxygenase, hydroxylase, glucosylase, and ketolase did not require in vitro evolution of each individual enzyme activity to achieve conversion of the new carotenoid substrates produced by these evolved pathways. Instead, these downstream enzymes were able to accept the related carotenoid substrates and to catalyze the synthesis of a number of structurally novel carotenoid compounds in *E. coli* (73). Further, the diversity of carotenoids produced in *E. coli* was increased by a carotenoid desaturase homolog identified from *S. aureus* that added oxygen-containing end groups to linear carotenoid molecules (10).

Directed evolution of the carotenoid synthases, the enzymes responsible for synthesizing the carotenoid backbone from either two FPP (C30 carotenoids) or GGDP (C40 carotenoids) molecules, has generated enzymes able to synthesize carotenoids of many different chain lengths. The C30 carotenoid synthase from *S. aureus* was evolved to produce C40 carotenoids (80) and further rounds of directed evolution yielded unnatural C45 and C50 carotenoids (81). In addition, C15 and C20 diphosphate precursors have been combined by a carotenoid synthase to form a novel, asymmetric C35 carotenoid (82).

2. Terpenes

Cyclization of linear prenyl chains by terpene cyclases yields an enormous array of structurally diverse terpene skeletons, and bioactive terpene natural products have been isolated from fungi, plants, and bacteria. Many terpenes of pharmaceutical and industrial interest are derived from plant secondary metabolism and, as a consequence, a number of plant terpene cyclases have been identified and characterized. However, despite the availability of several plant cyclases, production of bioactive plant terpenes in engineered microbes is often challenging because of nonoptimal codon usage and lack of information on enzymes that modify terpene scaffolds to produce the bioactive compound. Genomic information is limited for plants and, unlike bacteria and to some extent in fungi, plant biosynthetic genes are not clustered and many modifying enzymes are part of extensively duplicated enzyme families (e.g., P450 monooxygenases, glycosyltransferases), which makes their identification difficult. However, while this may limit full biosynthesis of some terpenes, it is possible to generate biologically active terpenes using semisynthetic strategies. For example, pathways that produce intermediates for the compounds taxol and artemisinin have been engineered in *E. coli* and yeast, and additional steps using traditional chemistry have completed the reaction sequence (76, 77).

Optimization and diversification of recombinant terpene biosynthesis benefit from both rational protein engineering and directed evolution of the terpene cyclases (1). A single terpene cyclase often has a broad product spectrum, producing a mixture of terpenes in different amounts from a single linear prenyl substrate. This heterogeneity is the result of a carbocation-initiated cyclization mechanism of these enzyme, where the cyclase controls migration of a reactive carbocation through the isoprene chain to produce a terminal carbocation that is finally quenched by a base (83). The "sloppiness" of a cyclase in controlling carbocation migration and quenching is reflected by its product spectrum. Most terpene cyclases have evolved to guide the cyclization preferentially to a limited number of products; however, only minimal changes are required to alter fundamentally the product spectrum of the reaction and relatively few amino acid residues may control the relative amounts of each terpene (84). For example, error-prone PCR and directed-evolution schemes have altered cadinene synthase to produce higher amounts of the related molecule germacrene (85) and rational engineering of the γ-humulene synthase yielded seven families of terpene sequences each able to produce specifically unique terpene scaffolds (86).

Subsequent reactions in terpene biosynthesis often are difficult to engineer into a heterologous host. Common to many terpene biosynthetic pathways, P450 enzymes that catalyze oxygenation steps often are not expressed well or require additional proteins able to transfer electrons from the reaction. Recently, several P450 enzymes from the taxol biosynthetic pathway have been functionally expressed in a heterologous system; however, at the same time, taxol pathway engineering has also stumbled at this point (77). Taxol, a diterpenoid isolated from the bark of the Pacific Yew tree, interferes with normal microtubule growth and is a potent drug in the treatment of several types of cancers. From the results of random sequencing of complementary deoxyribonucleic acids (cDNAs) produced during induction of taxol in *Taxus* cell culture, 19 enzymes are thought to catalyze the complete synthesis, and 8 of the enzymes are expected to be cytochrome P450 oxygenases (87). Five of the enzymes have been coexpressed in yeast resulting in mostly production of the taxadiene product of the terpene cyclase, while the subsequent P450 reaction generated only minimal amounts of the next intermediate (77). Because yeasts are able to express this P450 enzyme as a soluble and active protein, the hope remains that coexpressing the proper *Taxus* NADPH–cytochrome P450 reductase with the oxygenase will improve the activity of this reaction.

While the above example illustrates the difficulties of P450 expression, artemisinin production by both *E. coli* and yeast demonstrates that it is possible to obtain active P450 enzymes in heterologous expression hosts. Artemisinin, a sesquiterpene from *Artemisia annua*, is used to fight malarial infection and currently is extracted from the native plant. Pathway engineering efforts have developed both yeast and *E. coli* strains for the production of amorphodiene, the cyclized terpene scaffold that is further modified by subsequent oxidation and reduction reactions to artemisinin (74, 76). The production levels of amorphadiene by recombinant *E. coli* have been improved by engineering of an alternative isoprene precursor pathway (mevalonate pathway, see above) (88) and fermentation strategies to collect the volatile product (89). A breakthrough in extending the biosynthesis of amorphodiene to artemisinin occured with the cloning and functional expression of the P450 that catalyzes the three-step oxidation of amorphodiene to artemisinic acid (76). The key insight into cloning the P450 responsible for the oxidation of amorphodiene appears to be that several terpene hydroxylases from closely related plants belong to the CYP71 family of P450s. Coexpression of amorphodiene cyclase and oxidase produced up to 100 mg/L in vivo in yeast.

IV. Biosynthetic Pathway Optimization

The efficient conversion of simple sugars to complex chemicals by an engineered pathway requires optimization of the host's metabolism and enzyme activities directly involved in the biosynthetic pathway

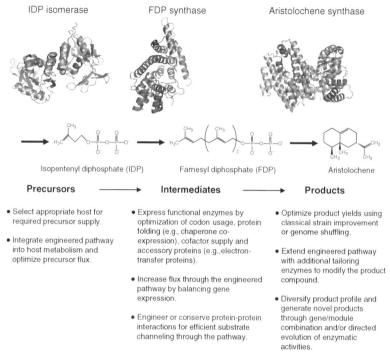

Precursors ⟶ Intermediates ⟶ Products

• Select appropriate host for required precursor supply. • Integrate engineered pathway into host metabolism and optimize precursor flux.	• Express functional enzymes by optimization of codon usage, protein folding (e.g., chaperone co-expression), cofactor supply and accessory proteins (e.g.,electron-transfer proteins). • Increase flux through the engineered pathway by balancing gene expression. • Engineer or conserve protein-protein interactions for efficient substrate channeling through the pathway.	• Optimize product yields using classical strain improvement or genome shuffling. • Extend engineered pathway with additional tailoring enzymes to modify the product compound. • Diversify product profile and generate novel products through gene/module combination and/or directed evolution of enzymatic activities.

Figure 3.   Representative pathway illustrates the considerations for engineering a biosynthetic reaction sequence. Aristolochene, a sesquiterpene, is the product of aristolochene synthase and is formed from isopentenyl diphosphate precursors. The enzymes for the biosynthesis of aristolochene are shown as ribbon diagrams and the substrate, intermediate, and product structures are shown. Potential problems with and solutions for engineering the pathway are listed.

(see Figure 3). Many potential problems can arise considering that the host strain may not have adapted to the presence of these particular enzymes or chemicals, and analogously the enzymes may react poorly within the environment of the host. The accumulation of intermediate compounds may occur because poor gene expression and protein folding limit the activities of individual enzymes or in novel pathways enzymes may be presented with nonideal substrates. The flux through an assembled recombinant pathway may be limited by the availability of precursor compounds or enzyme cofactors, and enzymes that have not evolved together to channel substrates to subsequent enzymes may reduce the output of the pathway. Taking all this into consideration, it is no surprise that pathway optimization strategies require manipulation of both the metabolic network of the host and individual recombinant biosynthetic genes.

Carotenoid biosynthesis in *E. coli* has been used as model systems to explore pathway optimization strategies because the colored pigments provide a convenient spectroscopic handle on the output of the pathway and intermediate and product compounds often can be distinguished easily. One of the first factors to consider during the optimization of a pathway is the expression level of the enzymes, including both total and relative expression levels. Addressing the total level of pathway expression, Kim et al. (70) determined that the araBAD promoter on medium-copy-number plasmids provided the highest production of lycopene in *E. coli*. Further studies have evaluated the relative levels of protein expression and have shown that significant contributions to product formation can be obtained simply by changing the activity of the rate-limiting enzyme. In a first example of this, directed evolution of geranylgeranyldiphosphate synthase, the first committed carotenoid enzyme, and its upstream promoter region identified mutations that increase carotenoid production (90). Following this idea, Alper et al. (91) measured lycopene production as a function of the relative strength between the promoters for deoxy-xylulose-P synthase (*dxs*) and the downstream enzymes (*ispD, ispF*) and isopentenyl pyrophosphate isomerase (*idi*) to determine the rate-limiting enzyme in lycopene biosynthesis. Specific promoters with calibrated expression strengths were integrated in front of the *dxs* gene to control in a continuous fashion its expression level, while the *ispD, ispF*, and *idi* genes were expressed either with the native promoter inducing low expression or a strong promoter for overexpression of these proteins. During low-level expression of *ispD, ispF*, and *idi*, lycopene production peaked at an optimal expression level for *dxs*, indicating that further increases of *dxs* levels are detrimental due to direct metabolic stresses related to protein expression or through accumulation of presumably toxic intermediate molecules. Alternatively, when *ispD, ispF*, and *idi* were overexpressed and the *dxs* promoter strength varied, lycopene production was proportional to the amount of *dxs* enzyme, suggesting that the rate-limiting step in the engineered pathway was catalyzed by *dxs*.

Methods for tuning the expression levels of several expressed enzymes have been developed using the nucleotide sequence between genes on a single messenger RNA (mRNA) transcript. Using a combinatorial library containing intergenic control elements that include mRNA secondary structures, RNase cleavage sites, and ribosome binding sequences, Pfleger et al. (92) generated changes in expression between two reporter genes that vary continuously over a 100-fold range, and applying this to the mevalonate pathway, balanced gene expression increased mevalonate production significantly (92).

Different strategies have been explored for optimizing the integration of a recombinant pathway into the metabolic network of a heterologous host. Control elements have been designed to couple heterologous pathway flux to the glycolytic flux of the host. For example, flux through an engineered carotenoid pathway in *E. coli* was controlled by cellular levels of pyruvate and glyceraldehyde-3-phosphate. Farmer et al. (93) designed a control element that stimulated expression of two rate-limiting carotenoid genes only during times of high substrate availability. To identify genes in *E. coli* that enhance lycopene-production when overexpressed, a genomic shotgun expression library of *E. coli* was transformed into lycopene-producing recombinant *E. coli* (94). Thirteen genes were identified that enhanced lycopene production; however, only one gene was directly involved in isoprenoid precursor biosynthesis, while the other genes were involved in regulating processes related to stationary-phase and anaerobic growth and may indirectly affect metabolic processes beneficial for carotenoid production. Other studies describe in silico metabolic modeling and transposon mutagenesis for the identification of genes in *E. coli* that when overexpressed or deleted will enhance lycopene production levels (95, 96).

Because of the complexity of metabolic networks, strategies involving the introduction of random mutations into the genome and screening for the desired phenotype continue to be successful approaches for the identification of improved metabolic traits. Classical strain improvement involves treatment of an initial strain with a mutagen followed by screening of the resulting phenotypes for the desired trait(s). Repeated cycles of mutation and selection can improve the characteristics of the strain significantly. Such strategies have been used to increase the resistance of a microbe to specific toxic conditions that are present during fermentation of a desired toxic product such as ethanol or phenol or are present because they are the target of enzymatic biodegradation of a toxic compound. To produce phenol from glucose and avoid the toxicity of phenolic aromatic compounds, Wierckx et al. (97) introduced a tyrosine phenol lyase into the solvent-resistant microbe *P. putida* S12 and used a combination of rational and random engineering and high-throughput screening to optimize production of phenol. Also using strain improvement strategies, Sonderegger et al. (98) developed a screen to select for cells able to continue to live without spending energy on replication and growth processes in ammonia- and glycerol-limited conditions. This strategy, to a large extent, decoupled the cell growth with production of an engineered chemical and produced a strain that retained metabolic vigor even when on the verge of starvation.

The selection strategy in strain improvements can be an arduous process and may take significant time to determine the improved strain. Additionally, many rounds of mutation and selection may be required to obtain a strain with the sufficiently altered phenotype, further increasing the length of the process. Whole genome-shuffling approaches have been developed to incorporate multiple changes in the genome by combining chemical mutagenesis with gene recombination techniques known as protoplast fusion. As in traditional strain improvement, the fittest strains are identified by a high-throughput screening strategy after being subjected to random mutagenesis; however, in whole-genome shuffling after only one round of selection the genomes of the fittest strains are shuffled to create strains that have combinations of the beneficial mutations. In general, after only a few rounds of shuffling the final strains have incorporated many beneficial changes and are significantly fitter than strains developed using classical strain improvement. In its first application, two rounds of genome shuffling matched over 20 years and 20 rounds of classical improvement methods in developing strains for the production of a complex polyketide antibiotic (99). Fermentation conditions and toxicity stresses have also been addressed by genome-shuffling methods. Applied to growth at low pH, five rounds of genome shuffling and selection for growth on plates with a pH gradient isolated *Lactobacillus* strains thriving at pH 4 that produce lactate at elevated levels (100). Degradation of the pesticide pentachlorophenol (PCP) by *Sphingobium chlorophenolicum* was improved by generation of strains that express the PCP-degrading enzymes constitutively and have developed resistance to the toxic effects of PCP (101). Recently, protoplast fusion in *E. coli* has been improved by a factor of 10^4 to present new avenues for evolutionary engineering in this commonly used host (102).

V.   Toward a Synthetic Biology

Developments in pathway engineering are leading to a synthetic biology by which microbial cell factories will produce novel and natural products from simple fermentations. General heterologous hosts such as *E. coli* and yeast will become robust production organisms, and strategies for expression of soluble and active protein will follow common protocols. As discussed in this review, methods for optimization of precursor supplies and balancing enzyme activity for increased carbon flux through the pathway are being designed. Also highlighted are the insights gained from structural biochemistry that define linker regions that connect enzyme modules specifically and efficiently to channel the substrate along the reaction sequence.

The efforts of systems biology will give shape to the construction and rational design of a host genome for production of specific compounds. The limiting precursors may be controlled by expression of particular pathways and gene knockouts or additional enzymes will be engineered. Theoretical and experimental studies will combine to provide a holistic view of the engineered pathway (103–105).

Many elements to control the regulation of biosynthetic pathways are in progress. The self-regulation of gene expression and enzyme activity in response to environmental conditions such as precursor supply is an important next step in the field of metabolic engineering (106, 107). A coupled set of three transcription factors has driven periodic expression of GFP (108), and transient chemical or thermal induction has been used to drive a bistable network between states using two repressible promoters (109). Using a random and modular construction, combinatorial synthesis of regulatory circuits using variations of promoters and transcription factors has been assembled to generate circuits with unique responses to environmental conditions (110). In an evolution of an engineered circuit, the promoter regions have been selected to match protein levels and used to convert a nonfunctional circuit into one that is functional, mainly by a change in the allosteric response of a truncated protein (111).

Standardization of the methods of synthetic biology may provide a set of biological parts that can be used for many applications simply by combining them in novel arrangements adapted for the specific purpose (107, 112). Component parts that behave in modular fashion for a specific function are the dream of pathway engineers, and as the PKS systems illustrate, there is precedent for this linked function. As synthetic DNA technologies and genomic databases mature and metabolic engineering gains more experience with biosynthetic pathways, we expect that designing new pathways will be a combination of gene selection, linker region design, and strain selection.

References

1. Chang, M. C. Y., and Keasling, J. D. (2006) Production of isoprenoid pharmaceuticals by engineered microbes, *Nat. Chem. Biol. 2*, 674–681.

2. Khosla, C., and Keasling, J. D. (2003) Timeline—Metabolic engineering for drug discovery and development, *Nat. Rev. Drug Discov.* 2, 1019–1025.

3. Petri, R., and Schmidt-Dannert, C. (2004) Dealing with complexity: Evolutionary engineering and genome shuffling, *Curr. Opin. in Biotechnol.* 15, 298–304.

4. Umeno, D., Tobias, A. V., and Arnold, F. H. (2005) Diversifying carotenoid biosynthetic pathways by directed evolution, *Microbiol. Mol. Biol. Rev.* 69, 51–78.

5. Koffas, M., and Stephanopoulos, G. (2005) Strain improvement by metabolic engineering: Lysine production as a case study for systems biology, *Curr. Opin. Biotechnol.* 16, 361–366.

6. Thykaer, J., and Nielsen, J. (2003) Metabolic engineering of beta-lactam production, *Met. Eng.* 5, 56–69.

7. Rusch, D. B., et al. (2007) The Sorcerer II Global Ocean Sampling expedition: Northwest Atlantic through Eastern Tropical Pacific, *PLOS Biol.* 5, 398–431.

8. Altschul, S. F., Gish, W., Miller, W., Myers, E. W., and Lipman, D. J. (1990) Basic Local Alignment Search Tool, *J. Mol. Biol.* 215, 403–410.

9. Peterson, J. D., Umayam, L. A., Dickinson, T., Hickey, E. K., and White, O. (2001) The Comprehensive Microbial Resource, *Nucl. Acids Res.* 29, 123–125.

10. Mijts, B. N., Lee, P. C., and Schmidt-Dannert, C. (2005) Identification of a carotenoid oxygenase synthesizing acyclic xanthophylls: Combinatorial biosynthesis and directed evolution, *Chem. Biol.* 12, 453–460.

11. Rontani, J. F., Bonin, P. C., and Volkman, J. K. (1999) Production of wax esters during aerobic growth of marine bacteria on isoprenoid compounds, *Appl. Environ. Microbiol.* 65, 221–230.

12. Kalscheuer, R., and Steinbuchel, A. (2003) A novel bifunctional wax ester synthase/acyl-CoA: Diacylglycerol acyltransferase mediates wax ester and triacylglycerol biosynthesis in Acinetobacter calcoaceticus ADP1, *J. Biol. Chem.* 278, 8075–8082.

13. Huu, N. B., Denner, E. B. M., Ha, D. T. C., Wanner, G., and Stan-Lotter, H. (1999) Marinobacter aquaeolei sp. nov., a halophilic bacterium isolated from a Vietnamese oil-producing well, *Int. J. Syst. Bacteriol.* 49, 367–375.

14. Holtzapple, E., and Schmidt-Dannert, C. (2007) Biosynthesis of isoprenoid wax ester in Marinobacter, hydrocarbonoclasticus DSM 8798: Identification and characterization of isoprenoid coenzyme A synthetase and wax ester synthases, *J. Bacteriol.* 189, 3804–3812.

15. Schmidt, E. W., Nelson, J. T., Rasko, D. A., Sudek, S., Eisen, J. A., Haygood, M. G., and Ravel, J. (2005) Patellamide A and C biosynthesis by a microcin-like pathway in Prochloron didemni, the cyanobacterial symbiont of Lissoclinum patella, *Proc. Nat. Acad. Sci. USA* 102, 7315–7320.

16. Kanehisa, M., Goto, S., Hattori, M., Aoki-Kinoshita, K. F., Itoh, M., Kawashima, S., Katayama, T., Araki, M., and Hirakawa, M. (2006) From genomics to chemical genomics: New developments in KEGG, *Nucl. Acids Res.* 34, D354–D357.

17. Karp, P. D., Ouzounis, C. A., Moore-Kochlacs, C., Goldovsky, L., Kaipa, P., Ahren, D., Tsoka, S., Darzentas, N., Kunin, V., and Lopez-Bigas, N. (2005) Expansion of the BioCyc collection of pathway/genome databases to 160 genomes, *Nucl. Acids Res.* 33, 6083–6089.

18. Keseler, I. M., Collado-Vides, J., Gama-Castro, S., Ingraham, J., Paley, S., Paulsen, I. T., Peralta-Gill, M., and Karp, P. D. (2005) EcoCyc: A comprehensive database resource for Escherichia coli, *Nucl. Acids Res. 33*, D334–D337.

19. Caspi, R., Foerster, H., Fulcher, C. A., Hopkinson, R., Ingraham, J., Kaipa, P., Krummenacker, M., Paley, S., Pick, J., Rhee, S. Y., Tissier, C., Zhang, P. F., and Karp, P. D. (2006) MetaCyc: A multiorganism database of metabolic pathways and enzymes, *Nucl. Acids Res. 34*, D511–D516.

20. Coates, J. D., Councell, T., Ellis, D. J., and Lovley, D. R. (1998) Carbohydrate oxidation coupled to Fe(III) reduction, a novel form of anaerobic metabolism, *Anaerobe 4*, 277–282.

21. Clardy, J., Fischbach, M. A., Walsh, C. T. (2006) New antibiotics from bacterial natural products, *Nat. Biotechnol. 24*, 1541–1550.

22. Watanabe, K., Hotta, K., Praseuth, A. P., Koketsu, K., Migita, A., Boddy, C. N., Wang, C. C. C., Oguri, H., and Oikawa, H. (2006) Total biosynthesis of antitumor nonribosomal peptides in Escherichia coli, *Nat. Chem. Biol. 2*, 423–428.

23. Watanabe, K., and Oikawa, H. (2007) Robust platform for de novo production of heterologous polyketides and nonribosomal peptides in Escherichia coli, *Org. Biomol. Chem. 5*, 593–602.

24. Sanchez, C., Zhu, L. L., Brana, A. F., Salas, A. P., Rohr, J., Mendez, C., and Salas, J. A. (2005) Combinatorial biosynthesis of antitumor indolocarbazole compounds, *Proc. Nat. Acad. Sci. USA 102*, 461–466.

25. Hahn, M., and Stachelhaus, T. (2006) Harnessing the potential of communication-mediating domains for the biocombinatorial synthesis of nonribosomal peptides, *Proc. Nat. Acad. Sci. USA 103*, 275–280.

26. Du, L. H., Sanchez, C., and Shen, B. (2001) Hybrid peptide-polyketide natural products: Biosynthesis and prospects toward engineering novel molecules, *Met. Eng. 3*, 78–95.

27. Leonard, E., Chemler, J., Lim, K. H., and Koffas, M. A. G. (2006) Expression of a soluble flavone synthase allows the biosynthesis of phytoestrogen derivatives in Escherichia coli, *Appl. Microbiol. Biotechnol. 70*, 85–91.

28. Watts, K. T., Lee, P. C., and Schmidt-Dannert, C. (2004) Exploring recombinant flavonoid biosynthesis in metabolically engineered Escherichia coli, *Chembiochem 5*, 500–507.

29. Leonard, E., Yang, Y. J., Chemler, J., Koffas, M. (2005) Metabolic engineering of flavonoid biosynthesis in Escherichia coli, *Abstr. Pap. Am. Chem. Soc. 229*, U189–U189.

30. Chemler, J., Yan, Y. J., and Koffas, M. A. G. (2006) Biosynthesis of isoprenoids, polyunsaturated fatty acids and flavonoids in Saccharomyces cerevisiae, *Microb. Cell Factories 5*, 20–29.

31. Kwon, S. J., de Boer, A. L., Petri, R., and Schmidt-Dannert, C. (2003) High-level production of porphyrins in metabolically engineered Escherichia coli: Systematic extension of a pathway assembled from overexpressed genes involved in heme biosynthesis, *Appl. Environ. Microbiol. 69*, 4875–4883.

32. Kwon, S. J., Petri, R., deBoer, A. L., and Schmidt-Dannert, C. (2004) A high-throughput screen for porphyrin metal chelatases: Application to the directed evolution of ferrochelatases for metalloporphyrin biosynthesis, *Chembiochem 5*, 1069–1074.

33. Cane, D. E., Walsh, C. T., and Khosla, C. (1998) Biochemistry—Harnessing the biosynthetic code: Combinations, permutations, and mutations, *Science 282*, 63–68.

34. Baltz, R. H. (2006) Molecular engineering approaches to peptide, polyketide and other antibiotics, *Nat. Biotechnol. 24*, 1533–1540.

35. Khosla, C., Tang, Y., Chen, A. Y., Schnarr, N. A., and Cane, D. E. (2007) Structure and mechanism of the 6-deoxyerythronolide B synthase, *Annu. Rev. Biochem. 76*, 195–221.

36. Walsh, C. T. (2004) Polyketide and nonribosomal peptide antibiotics: Modularity and versatility, *Science 303*, 1805–1810.

37. Zhang, C. S., Griffith, B. R., Fu, Q., Albermann, C., Fu, X., Lee, I. K., Li, L. J., and Thorson, J. S. (2006) Exploiting the reversibility of natural product glycosyltransferase-catalyzed reactions, *Science 313*, 1291–1294.

38. Vaillancourt, F. H., Yeh, E., Vosburg, D. A., Garneau-Tsodikova, S., and Walsh, C. T. (2006) Nature's inventory of halogenation catalysts: Oxidative strategies predominate, *Chem. Rev. 106*, 3364–3378.

39. Chan, Y. A., Boyne, M. T., Podevels, A. M., Klimowicz, A. K., Handelsman, J., Kelleher, N. L., and Thomas, M. G. (2006) Hydroxymalonyl-acyl carrier protein (ACP) and aminomalonyl-ACP are two additional type I polyketide synthase extender units, *Proc. Nat. Acad. Sci. USA 103*, 14349–14354.

40. Yin, J., Straight, P. D., Hrvatin, S., Dorrestein, P. C., Bumpus, S. B., Jao, C., Kelleher, N. L., Kolter, R., and Walsh, C. T. (2007) Genome-wide high-throughput mining of natural-product biosynthetic gene clusters by phage display, *Chem. Biol. 14*, 303–312.

41. Pfeifer, B. A., Admiraal, S. J., Gramajo, H., Cane, D. E., and Khosla, C. (2001) Biosynthesis of complex polyketides in a metabolically engineered strain of E-coli, *Science 291*, 1790–1792.

42. Pfeifer, B., Hu, Z. H., Licari, P., and Khosla, C. (2002) Process and metabolic strategies for improved production of Escherichia coli-derived 6-deoxyerythronolide B, *Appl. Environ. Microbiol. 68*, 3287–3292.

43. Mutka, S. C., Bondi, S. M., Carney, J. R., Da Silva, N. A., and Kealey, J. T. (2006) Metabolic pathway engineering for complex polyketide biosynthesis in Saccharomyces cerevisiae, *FEMS Yeast Res. 6*, 40–47.

44. Gross, F., Luniak, N., Perlova, O., Gaitatzis, N., Jenke-Kodama, H., Gerth, K., Gottschalk, D., Dittmann, E., and Muller, R. (2006) Bacterial type III polyketide synthases: Phylogenetic analysis and potential for the production of novel secondary metabolites by heterologous expression in pseudomonads, *Arch. Microbiol. 185*, 28–38.

45. Menzella, H. G., Reisinger, S. J., Welch, M., Kealey, J. T., Kennedy, J., Reid, R., Tran, C. Q., and Santi, D. V. (2006) Redesign, synthesis and functional expression of the 6-deoxyerythronolide B polyketide synthase gene cluster, *J. Ind. Microbiol. Biotechnol. 33*, 22–28.

46. Lee, H. Y., and Khosla, C. (2007) Bioassay-guided evolution of glycosylated macrolide antibiotics in Escherichia coli, *PLOS Biol. 5*, 243–250.

47. Stephan, S., Heinzle, E., Wenzel, S. C., Krug, D., Muller, R., and Wittmann, C. (2006) Metabolic physiology of Pseudomonas putida for heterologous production of myxochromide, *Process Biochem. 41*, 2146–2152.

48. Wenzel, S. C., Gross, F., Zhang, Y., Fu, J., Stewart, A. F., and Muller, R. (2005) Heterologous expression of a myxobacterial natural products assembly line in pseudomonads via red/ET recombineering, *Chem. Biol. 12*, 349–356.

49. Wenzel, S. C., and Muller, R. (2005) Recent developments towards the heterologous expression of complex bacterial natural product biosynthetic pathways, *Curr. Opin. Biotechnol. 16*, 594–606.

50. Peiru, S., Menzella, H. G., Rodriguez, E., Carney, J., and Gramajo, H. (2005) Production of the potent antibacterial polyketide erythromycin C in Escherichia coli, *Appl. Environ. Microbiol. 71*, 2539–2547.

51. Perez, M., Lombo, F., Baig, I., Brana, A. F., Rohr, J., Salas, J. A., and Mendez, C. (2006) Combinatorial biosynthesis of antitumor deoxysugar pathways in Streptomyces griseus: Reconstitution of "unnatural natural gene clusters" for the biosynthesis of four 2,6-D-dideoxyhexoses, *Appl. Environ. Microbiol. 72*, 6644–6652.

52. Luzhetskyy, A., and Bechthold, A. (2005) It works: Combinatorial biosynthesis for generating novel glycosylated compounds, *Mol. Microbiol. 58*, 3–5.

53. Hill, A. M. (2006) The biosynthesis, molecular genetics and enzymology of the polyketide-derived metabolites, *Nat. Prod. Rep. 23*, 256–320.

54. Weissman, K. J., and Leadlay, P. F. (2005) Combinatorial biosynthesis of reduced polyketides, *Nat. Rev. Microbiol. 3*, 925–936.

55. McDaniel, R., Welch, M., and Hutchinson, C. R. (2005) Genetic approaches to polyketide antibiotics. 1, *Chem. Rev. 105*, 543–558.

56. McDaniel, R., Thamchaipenet, A., Gustafsson, C., Fu, H., Betlach, M., Betlach, M., and Ashley, G. (1999) Multiple genetic modifications of the erythromycin polyketide synthase to produce a library of novel "unnatural" natural products, *Proc. Nat. Acad. Sci. USA 96*, 1846–1851.

57. Menzella, H. G., Reid, R., Carney, J. R., Chandran, S. S., Reisinger, S. J., Patel, K. G., Hopwood, D. A., and Santi, D. V. (2005) Combinatorial polyketide biosynthesis by de novo design and rearrangement of modular polyketide synthase genes, *Nat. Biotechnol. 23*, 1171–1176.

58. Kim, B. S., Sherman, D. H., and Reynolds, K. A. (2004) An efficient method for creation and functional analysis of libraries of hybrid type I polyketide synthases, *Prot. Eng. Design Selection 17*, 277–284.

59. Kittendorf, J. D., and Sherman, D. H. (2006) Developing tools for engineering hybrid polyketide synthetic pathways, *Curr. Opin. Biotechnol. 17*, 597–605.

60. Floss, H. G. (2006) Combinatorial biosynthesis—Potential and problems, *J. Biotechnol. 124*, 242–257.

61. Watanabe, K., Wang, C. C. C., Boddy, C. N., Cane, D. E., and Khosla, C. (2003) Understanding substrate specificity of polyketide synthase modules by generating hybrid multimodular synthases, *J. Biol. Chem. 278*, 42020–42026.

62. Schnarr, N. A., Chen, A. Y., Cane, D. E., and Khosla, C. (2005) Analysis of covalently bound polyketide intermediates on 6-deoxyerythronolide B synthase by tandem proteolysis-mass spectrometry, *Biochemistry 44*, 11836–11842.

63. Lai, J. R., Fischbach, M. A., Liu, D. R., and Walsh, C. T. (2006) Localized protein interaction surfaces on the EntB carrier protein revealed by combinatorial mutagenesis and selection, *J. Am. Chem. Soc. 128*, 11002–11003.

64. Lai, J. R., Koglin, A., and Walsh, C. T. (2006) Carrier protein structure and recognition in polyketide and nonribosomal peptide biosynthesis, *Biochemistry 45*, 14869–14879.

65. Schnarr, N. A., and Khosla, C. (2006) Trapping transient protein-protein interactions in polyketide biosynthesis, *ACS Chem. Biol. 1*, 679–680.

66. Worthington, A. S., Rivera, H., Torpey, J. W., Alexander, M. D., and Burkart, M. D. (2006) Mechanism-based protein cross-linking probes to investigate carrier protein-mediated biosynthesis, *ACS Chem. Biol. 1*, 687–691.

67. Gokhale, R. S., Tsuji, S. Y., Cane, D. E., and Khosla, C. (1999) Dissecting and exploiting intermodular communication in polyketide synthases, *Science 284*, 482–485.

68. Chandran, S. S., Menzella, H. G., Carney, J. R., and Santi, D. V. (2006) Activating hybrid modular interfaces in synthetic polyketide synthases by cassette replacement of keto-synthase domains, *Chem. Biol. 13*, 469–474.

69. Weissman, K. J. (2006) The structural basis for docking in modular polyketide biosynthesis, *Chembiochem 7*, 485–494.

70. Kim, S. W., and Keasling, J. D. (2001) Metabolic engineering of the nonmevalonate isopentenyl diphosphate synthesis pathway in Escherichia coli enhances lycopene production, *Biotechnol. Bioeng. 72*, 408–415.

71. Farmer, W. R., and Liao, J. C. (2001) Precursor balancing for metabolic engineering of lycopene production in Escherichia coli, *Biotechnol. Prog. 17*, 57–61.

72. Yuan, L. Z., Rouviere, P. E., LaRossa, R. A., and Suh, W. (2006) Chromosomal promoter replacement of the isoprenoid pathway for enhancing carotenoid production in E. coli, *Met. Eng. 8*, 79–90.

73. Lee, P. C., Momen, A. Z. R., Mijts, B. N., and Schmidt-Dannert, C. (2003) Biosynthesis of structurally novel carotenoids in Escherichia coli, *Chem. Biol. 10*, 453–462.

74. Martin, V. J. J., Pitera, D. J., Withers, S. T., Newman, J. D., and Keasling, J. D. (2003) Engineering a mevalonate pathway in Escherichia coli for production of terpenoids, *Nat. Biotechnol. 21*, 796–802.

75. Pitera, D. J., Paddon, C. J., Newman, J. D., and Keasling, J. D. (2007) Balancing a heterologous mevalonate pathway for improved isoprenoid production in Escherichia coli, *Met. Eng. 9*, 193–207.

76. Ro, D. K., Paradise, E. M., Ouellet, M., Fisher, K. J., Newman, K. L., Ndungu, J. M., Ho, K. A., Eachus, R. A., Ham, T. S., Kirby, J., Chang, M. C. Y., Withers, S. T., Shiba, Y., Sarpong, R., and Keasling, J. D. (2006) Production of the antimalarial drug precursor artemisinic acid in engineered yeast, *Nature 440*, 940–943.

77. DeJong, J. M., Liu, Y. L., Bollon, A. P., Long, R. M., Jennewein, S., Williams, D., and Croteau, R. B. (2006) Genetic engineering of Taxol biosynthetic genes in Saccharomyces cerevisiae, *Biotechnol. Bioeng. 93*, 212–224.

78. Schmidt-Dannert, C., Umeno, D., and Arnold, F. H. (2000) Molecular breeding of carotenoid biosynthetic pathways, *Nat. Biotechnol. 18*, 750–753.

79. Wang, C. W., and Liao, J. C. (2001) Alteration of product specificity of Rhodobacter sphaeroides phytoene desaturase by directed evolution, *J. Biol. Chem. 276*, 41161–41164.

80. Umeno, D., Tobias, A. V., and Arnold, F. H. (2002) Evolution of the C-30 carotenoid synthase CrtM for function in aC(40) pathway, *J. of Bacteriol. 184*, 6690–6699.

81. Umeno, D., and Arnold, F. H. (2004) Evolution of a pathway to novel long-chain carotenoids, *J. Bacteriol. 186*, 1531–1536.

82. Umeno, D., and Arnold, F. H. (2003) A C-35 carotenoid biosynthetic pathway, *Appl. Environ. Microbiol. 69*, 3573–3579.

83. Segura, M. J. R., Jackson, B. E., and Matsuda, S. P. T. (2003) Mutagenesis approaches to deduce structure-function relationships in terpene synthases, *Nat. Prod. Rep. 20*, 304–317.

84. Greenhagen, B. T., O'Maille, P. E., Noel, J. P., and Chappell, J. (2006) Identifying and manipulating structural determinates linking catalytic specificities in terpene synthases, *Proc. Nat. Acad. Sci. USA 103*, 9826–9831.

85. Yoshikuni, Y., Martin, V. J. J., Ferrin, T. E., and Keasling, J. D. (2006) Engineering cotton (+)-delta-cadinene synthase to an altered function: Germacrene D-4-ol synthase, *Chem. Biol. 13*, 91–98.

86. Yoshikuni, Y., Ferrin, T. E., and Keasling, J. D. (2006) Designed divergent evolution of enzyme function, *Nature 440*, 1078–1082.

87. Jennewein, S., Wildung, M. R., Chau, M., Walker, K., and Croteau, R. (2004) Random sequencing of an induced Taxus cell cDNA library for identification of clones involved in Taxol biosynthesis, *Proc. Nat. Acad. Sci. USA 101*, 9149–9154.

88. Shiba, Y., Paradise, E. M., Kirby, J., Ro, D. K., and Keasing, J. D. (2007) Engineering of the pyruvate dehydrogenase bypass in Saccharomyces cerevisiae for high-level production of isoprenoids, *Met. Eng. 9*, 160–168.

89. Newman, J. D., Marshall, J., Chang, M., Nowroozi, F., Paradise, E., Pitera, D., Newman, K. L., and Keasling, J. D. (2006) High-level production of amorpha-4,11-diene in a two-phase partitioning bioreactor of metabolically, engineered Escherichia coli, *Biotechnol. Bioeng. 95*, 684–691.

90. Wang, C. W., Oh, M. K., and Liao, J. C. (2000) Directed evolution of metabolically engineered Escherichia coli for carotenoid production, *Biotechnol. Prog. 16*, 922–926.

91. Alper, H., Fischer, C., Nevoigt, E., and Stephanopoulos, G. (2006) Tuning genetic control through promoter engineering, *Proc. Nat. Acad. Sci. USA 102*, 12678–12683.

92. Pfleger, B. F., Pitera, D. J., Smolke, C. D., and Keasling, J. D. (2006) Combinatorial engineering of intergenic regions in operons tunes expression of multiple genes, *Nat. Biotechnol. 24*, 1027–1032.

93. Farmer, W. R., and Liao, J. C. (2000) Improving lycopene production in Escherichia coli by engineering metabolic control, *Nat. Biotechnol. 18*, 533–537.

94. Jang, J. K., Pham, T. H., Chang, I. S., Kang, K. H., Moon, H., Cho, K. S., and Kim, B. H. (2004) Construction and operation of a novel mediator- and membrane-less microbial fuel cell, *Process Biochem. 39*, 1007–1012.

95. Alper, H., Miyaoku, K., and Stephanopoulos, G. (2005) Construction of lycopene-overproducing E-coli strains by combining systematic and combinatorial gene knockout targets, *Nat. Biotechnol. 23*, 612–616.

96. Alper, H., Jin, Y. S., Moxley, J. F., and Stephanopoulos, G. (2005) Identifying gene targets for the metabolic engineering of lycopene biosynthesis in Escherichia coli, *Met. Eng. 7*, 155–164.

97. Wierckx, N. J. P., Ballerstedt, H., de Bont, J. A. M., and Wery, J. (2005) Engineering of solvent-tolerant Pseudomonas putida S12 for bioproduction of phenol from glucose, *Appl. Environ. Microbiol. 71*, 8221–8227.

98. Sonderegger, M., Schumperli, M., and Sauer, U. (2005) Selection of quiescent Escherichia coli with high metabolic activity, *Met. Eng. 7*, 4–9.

99. Zhang, S. G. (2002) Emerging biological materials through molecular self-assembly, *Biotechnol. Adv. 20*, 321–339.

100. Patnaik, R., Louie, S., Gavrilovic, V., Perry, K., Stemmer, W. P. C., Ryan, C. M., and del Cardayre, S. (2002) Genome shuffling of Lactobacillus for improved acid tolerance, *Nat. Biotechnol. 20*, 707–712.

101. Dai, M. H., and Copley, S. D. (2004) Genome shuffling improves degradation of the anthropogenic pesticide pentachlorophenol by Sphingobium chlorophenolicum ATCC 39723, *Appl. Environ. Microbiol. 70*, 2391–2397.

102. Dai, M. H., Ziesman, S., Ratcliffe, T., Gill, R. T., and Copley, S. D. (2005) Visualization of protoplast fusion and quantitation of recombination in fused protoplasts of auxotrophic strains of Escherichia coli, *Met. Eng. 7*, 45–52.

103. Alper, H., and Stephanopoulos, G. (2004) Metabolic engineering challenges in the post-genomic era, *Chem. Eng. Sci. 59*, 5009–5017.

104. Stephanopoulos, G., Alper, H., and Moxley, J. (2004) Exploiting biological complexity for strain improvement through systems biology, *Nat. Biotechnol. 22*, 1261–1267.

105. Trinh, C. T., Carlson, R., Wlaschin, A., and Srienc, F. (2006) Design, construction and performance of the most efficient biomass producing E-coli bacterium, *Met. Eng. 8*, 628–638.

106. Chin, J. W. (2006) Programming and engineering biological networks, *Curr. Opin. Struct. Biol. 16*, 551–556.

107. Endy, D. (2005) Foundations for engineering biology, *Nature 438*, 449–453.

108. Elowitz, M. B., and Leibler, S. (2000) A synthetic oscillatory network of transcriptional regulators, *Nature 403*, 335–338.

109. Gardner, T. S., Cantor, C. R., and Collins, J. J. (2000) Construction of a genetic toggle switch in Escherichia coli, *Nature 403*, 339–342.

110. Guet, C. C., Elowitz, M. B., Hsing, W. H., and Leibler, S. (2002) Combinatorial synthesis of genetic networks, *Science 296*, 1466–1470.

111. Yokobayashi, Y., Weiss, R., and Arnold, F. H. (2002) Directed evolution of a genetic circuit, *Proc. Nat. Acad. Sci. USA 99*, 16587–16591.

112. Drubin, D. A., Way, J. C., and Silver, P. A. (2007) Designing biological systems, *Genes Dev. 21*, 242–254.

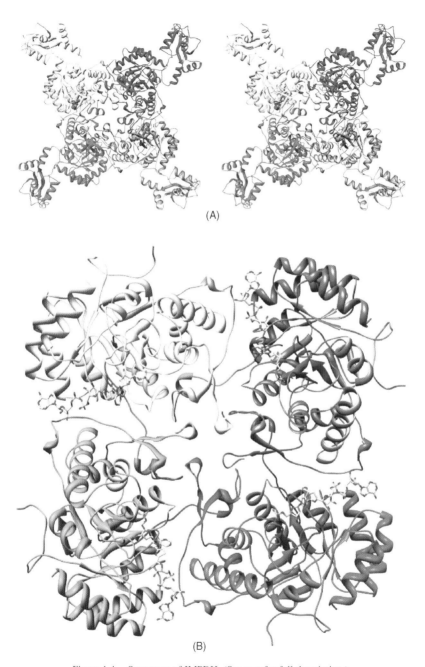

Figure 1.4.   Structures of IMPDH. (See text for full description.)

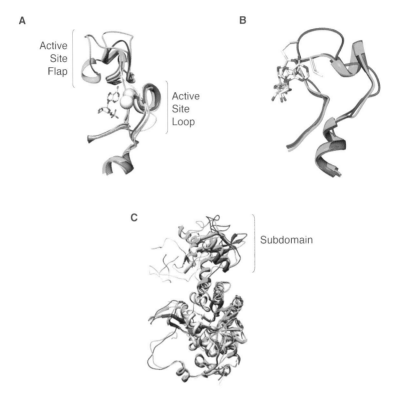

Figure 1.6.   IMPDH conformational flexibility. Superposition of α-carbon traces of monomer core domains in different structures. (A) Structural flexibility of the active-site loop and active-site flap. Superimposition of the entire core domains of 1jr1, 1b3o, 1eep, 1zfj, 1ak5, and 1pvn. The substrate IMP of *S. pyogenes* structure (1zfj) is shown in CPK-colored stick representation. The 1zfj active-site Cys is shown as a CPK-colored space-fill model. (B) Change in the active-site loop 6 conformation upon 6-Cl-IMP binding. Superimposition of the entire core domains of 1nfb, 1zfj, and 1jr1. The IMP substrate of 1zfj and the 6-Cl-IMP adduct are shown in stick representation. (C) Position of the CBS subdomain in different IMPDH structures relative to the enzyme core domain. The substrate IMP of 1zfj structure is illustrated in red space-fill representation. The human type 2 structure is in gold (1b3o), the hamster structure in dark blue (1jr1), the *S. pyogenes* (1zfj) structure in green, the *B. burgdorferi* structure (1eep) in grey, the *T. foetus* apo enzyme structure in magenta (1ak5), the human type 2 structure in complex with 6-Cl-IMP and NAD in red, and the *T. foetus* E·MMP complex in cyan (1pvn). UCSF Chimera (36a) was used for the coordinate superposition and structure visualization.

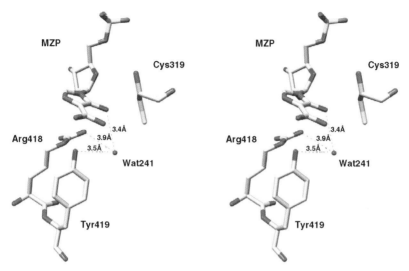

Figure 1.7. Stereoview of the transition state analogy of MMP in *T. foetus* structure (1pvn) and mechanism of water activation. The distances are indicated with dotted lines.

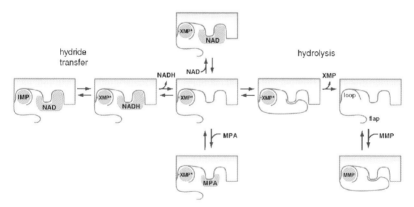

Figure 1.9. Plausible structural dynamics of the IMPDH catalytic cycle. The active-site loop 6 and the flap appear to be relatively disordered in the absence of substrates. Binding of IMP causes a conformational change in the loop; the flap remains largely disordered. After the hydride transfer is complete, NADH dissociates and the flap moves into the NAD site, activating water and converting the enzyme to a hydrolase.

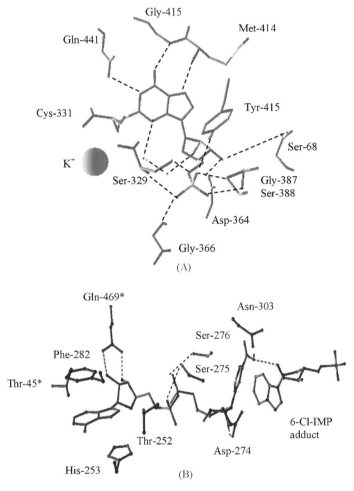

Figure 1.10. Structures of the active site. (A) The substrate site as seen in the hamster E-XMP*·MPA structure; the red spheres are water molecules (28). (B) SAD bound to the NAD site as seen in the complex with the human type 2 6-Cl-IMP adduct (31). (C) The potassium binding site as observed in the hamster IMPDH. Residues marked by asterisks are from a second subunit (28). Figure made using the program DeepView (95a).

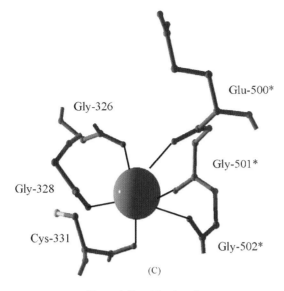

(C)

Figure 1.10. (*Continued*)

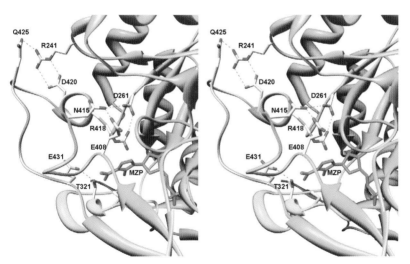

Figure 1.11.   Stereoview of the interactions of the flap with the NAD binding site in the closed enzyme conformation (*T. foetus* E·MMP complex; PDB file 1pvn). MZP is shown in magenta sticks. The distal flap (residues 413–431) is orange, the active-site loop 6 (residues 313–328) with the active-site Cys is cyan. Hydrogen bonds are shown as dotted lines.

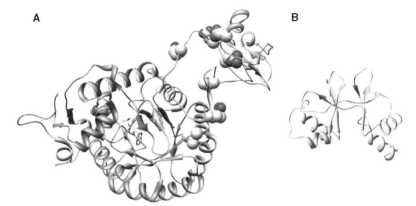

Figure 1.12. Subdomain (Bateman domain) of IMPDH. (A) Structure of human IMPDH type 1 (PDB code 1jcn) in complex with 6-Cl-IMP. The RP10-associated amino acid substitutions are shown in space-filling representation. The substrate is shown as sticks. (B) Structure of the Bateman domain of *S. pyogenes* IMPDH (PDB code 1zfj) (IMPDH residues 96–221). UCSF Chimera was used for structure visualization (36a).

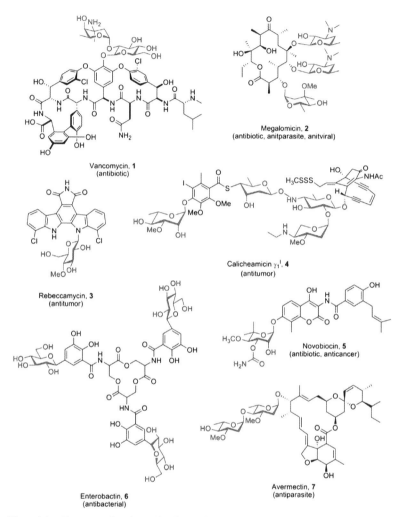

Figure 2.1.   Representative glycosylated natural products and their biological activities. These examples include *O*-glycosides (**1**, **2**, **4**, **5**, and **7**), *N*-glycosides (**3** and **4**), and a *C*-glycoside (**6**). The glycosyl moieties are highlighted blue.

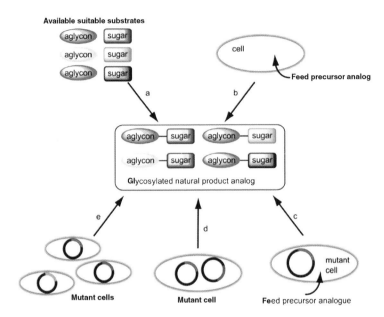

Figure 2.2. Strategies for glycodiversification of natural products. (See text for full description.)

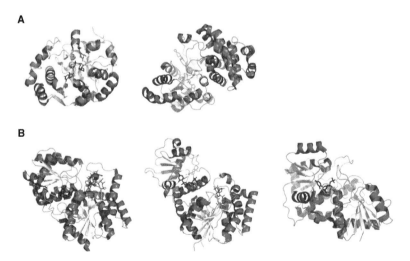

Figure 2.3. Structures of representative GTs. (See text for full description.)

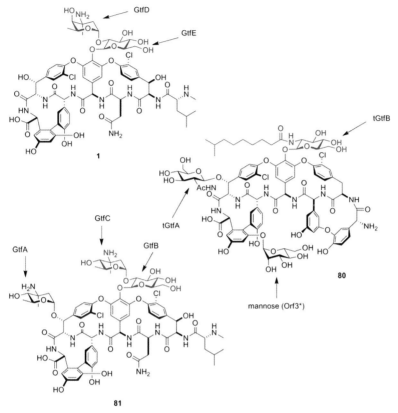

Figure 2.15. Naturally occurring glycopeptide antibiotics. Sugar moieties are highlighted blue.

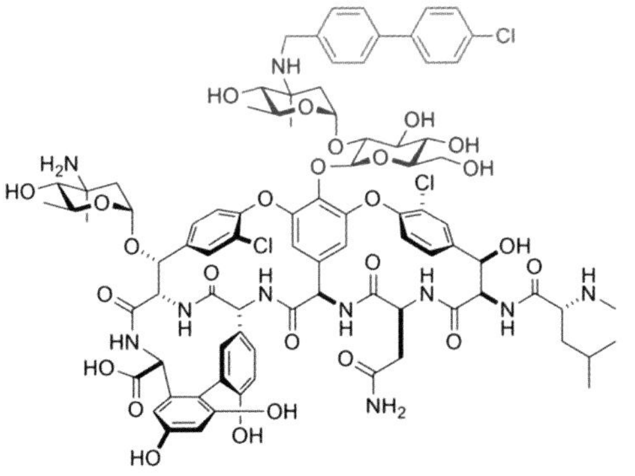

82

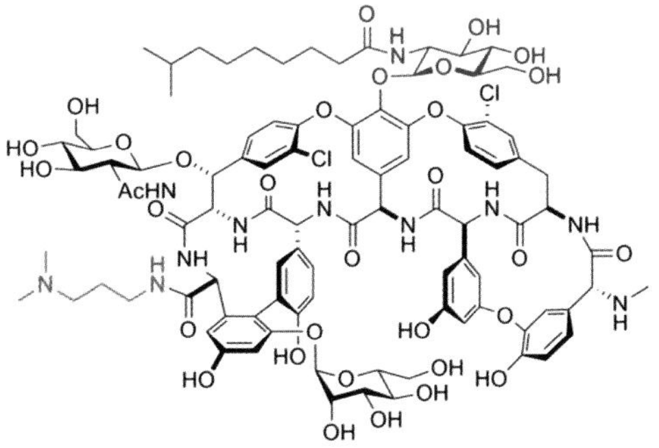

83

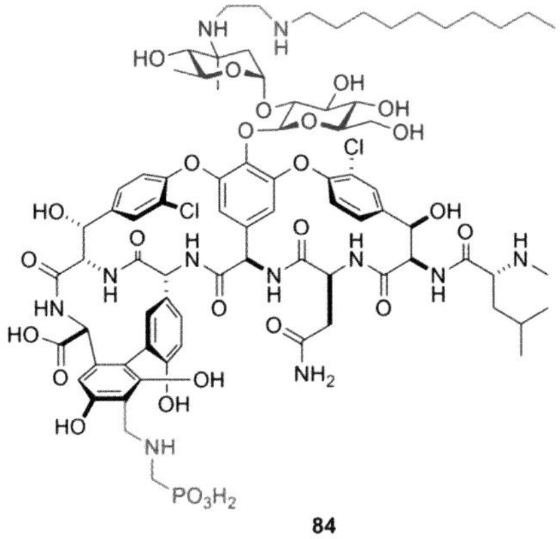

84

Figure 2.16.   Second-generation glycopeptide antibiotics. Sugar moieties are highlighted blue and structural deviations from the parent natural glycopeptide highlighted red.

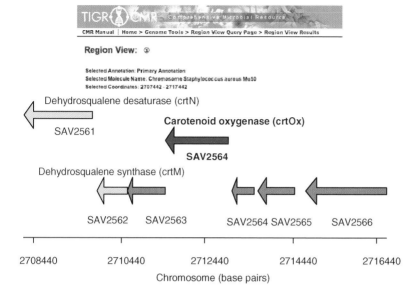

CMR Manual | Home > Genome Tools > Region View Query Page > Region View Results

Region View: ④

Selected Annotation: Primary Annotation
Selected Molecule Name: Chromosome Staphylococcus aureus Mu50
Selected Coordinates: 2707442 - 2717442

Dehydrosqualene desaturase (crtN)

SAV2561

Carotenoid oxygenase (crtOx)

SAV2564

Dehydrosqualene synthase (crtM)

SAV2562   SAV2563   SAV2564 SAV2565   SAV2566

2708440   2710440   2712440   2714440   2716440

Chromosome (base pairs)

Figure 3.1. Region of the *S. aureus* Mu50 genome surrounding several carotenoid genes illustrates the regional display tool from the TIGR CMR Web tool for observing neighboring ORFs. Arrows correspond to ORFs labeled with their gene identifications. The symbols *crtM* and *crtN* (gold arrows) are previously characterized carotenoid genes from *S. aureus* known to make diapolycopene (C30) carotenoids. A role for *crtOx* (violet arrow) in the carotenoid pathway was suggested by its proximity to *crtM* and *crtN* and subsequent biochemical characterization has identified CrtOx as a carotenoid oxygenase. Putative or hypothetical ORFs (gray arrows) are also shown.

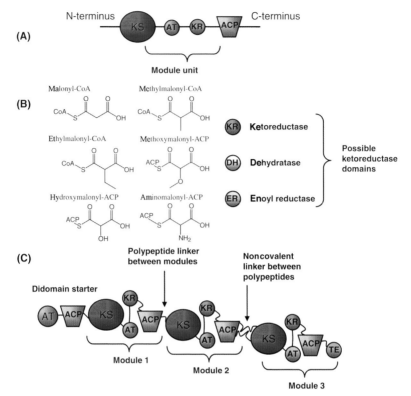

Figure 3.2. Type I polyketide synthases as an example of combinatorial biosynthesis. (A) The primary sequence of the multidomain enzyme complex contains several modules each possessing ketosynthase (KS), acyltransferase (AT), and ketoreductase (KR) domains and the initial and final modules containing starter and thioesterase (TE) domains. (B) Several types of extender units and KR domain catalytic activity define the reaction for each module unit. Engineering within each module has led to novel products and reaction sequences. (C) Several polypeptides interact to form the tertiary and quaternary structures that determine the complete reaction sequence. The linker regions that are defined by protein–protein interactions between consecutive modules are key structural elements that may be engineered to reorder reaction sequences for novel polyketides.

TRENDS IN MICROBIAL SYNTHESIS OF NATURAL PRODUCTS AND BIOFUELS

By JOSEPH A. CHEMIER, ZACHARY L. FOWLER, and MATTHEOS A. G. KOFFAS, *Department of Chemical and Biological Engineering, University at Buffalo, Buffalo, New York* and EFFENDI LEONARD, *Department of Chemical Engineering, Massachusetts Institute of Technology, Amherst, Massachusetts*

CONTENTS

Advances in Enzymology and Related Areas of Molecular Biology, Volume 76
Edited by Eric J. Toone   Copyright © 2009 by John Wiley & Sons, Inc.

I. Introduction

Traditional chemical manufacturing relies on the consumption of unsustainable resources for fuel and raw materials, and as a consequence the rapid depletion of natural resources has raised political, environmental, and social concerns on the sustainability of current manufacturing processes. Green chemistry presents a promising solution to this dilemma; however, the applicability and delivery have yet to be fully implemented. Biocatalyst-based processes are notably more advantageous than chemical synthesis for various reasons. For example, unlike chemical synthesis, enzyme-based reactions are generally performed under moderate temperatures and hence they limit energy consumption. The performance of enzymatic reactions also seldom requires the use of toxic solvents, which translates to reduced waste management, not to mention the specificity of enzymatic catalysis allows for improved product purity. Today, with the wealth of genetic information and advances in genetic engineering, bioinformatics, and systems biology, biochemical synthesis has taken a major leap toward the development of the microbial "factory". This chapter covers this emerging niche of biotechnology that has grown to become an important green chemistry curriculum to synthesize high-value chemicals and fuels.

II. Terpenoids

Terpenoids (also referred to as isoprenoids) are one of the most diverse groups of secondary metabolites with more than 55,000 molecules identified (1). Some of these compounds are classified as primary metabolites required for cellular maintenance and function but also serve functions for plant photoprotection and elongation (gibberellins, carotenoids, and sterols) while the remaining are classified as secondary metabolites (2). It is these secondary metabolic terpenoids that are of particular commercial interest as they are used in a wide array of applications ranging from flavors, fragrances, and cleaning products to anticancer and antimicrobial agents and potential pharmaceuticals (3). Total biosynthesis of some terpenoids such as ingenol (4), resiniferatoxin (5), and guanacastepenes (6) has been reported and can be achieved through a series of isoprene condensations and cyclizations. However, plant extraction is still the primary route to obtaining many terpenoids (7) despite low yields and its expensive nature. Therefore alternative routes that rely on plant and tissue cell cultures and recombinant microorganisms have been exploited for terpenoid biosynthesis to increase availability and product specificity (8, 9).

All terpenoids are derived from two universal precursors (referred to as isoprene units), namely isopentenyl pyrophosphate (IPP) and dimethylallyl pyrophosphate (DMAPP). These are derived from two distinct biosynthetic routes. The first one, known as the mevalonate (MVA) pathway, uses seven enzymes to supply the precursors in most eukaryotes, archaea, bacteria, the cytosol and mitochondria of plants, fungi, and *Trypanosoma* and *Leishmania* (10). The regulation of this pathway is known to be complex in all eukaryotic organisms examined (11, 12) and involves multiple levels of feedback inhibition. This feedback regulation includes changes in transcription, translation, and protein stability but also depends on the availability of molecular oxygen (13, 14).

The first step in the MVA pathway (Figure 1A) is the synthesis of acetoacetyl-CoA from two molecules of acetyl-CoA in a reaction catalyzed by the acetoacetyl-CoA thiolase (15) encoded by *ERG10* in yeast. In yeast, this enzyme is subjected to regulation by intracellular levels of sterols. The next step involves the condensation of acetyl-CoA with acetoacetyl-CoA to yield 3-hydroxy-3-methylglutaryl-CoA (HMG-CoA) which is catalyzed by HMG-CoA synthase, an enzyme with yet-unknown regulatory mechanism (16). The third step is catalyzed by the enzyme HMG-CoA reductase, leading to the biosynthesis of MVA. Mevalonate kinase phosphorylates MVA in the fourth step where its activity is feedback inhibited by farnesyl diphosphate (FPP) and geranyl diphosphate (GPP) (17). Catalyzed by phosphomevalonate kinase and encoded by *ERG8* in yeast, the next step leads to the synthesis of diphosphomevalonate. IPP is finally formed by diphosphatemevalonate decarboxylase, an enzyme encoded by *MVD1* in yeast. IPP is interconverted to DMAPP by isopentenyl diphosphate–dimethylallyl diphosphate isomerase, encoded by *IDI1*.

Relatively recently, labeling experiments in bacteria and plants revealed the presence of an alternative terpenoid biosynthetic pathway, named the MEP (Figure 2) pathway after its first committed precursor, 2-*C*-methyl-D-erythritol-4-phosphate (MEP) (18). It now appears that some plants, gram-negative bacteria, and *Bacillus subtilis* use the MEP pathway (19). In the first step of this pathway, 1-deoxy-D-xylulose 5-phosphate (DXP) synthase catalyzes the condensation of D-glyceraldehyde 3-phosphate and pyruvate to form DXP. The next step is the actual first committed step in the terpenoid biosynthesis where biocatalysis forms MEP by the reversible enzyme DXP

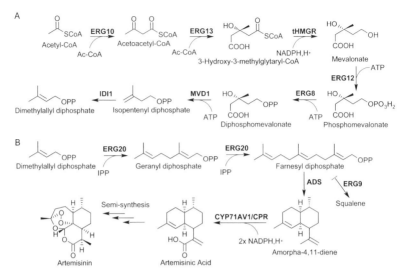

Figure 1.   (A) Mevalonate pathway leading to the formation of isopentyl diphosphate and dimethylallyl diphosphate. (B) Artemisinic acid pathway leading to the synthesis of artemisinic acid and artemisinin.   IPP: isopentenyl pyrophosphate; ERG10: acetoacetyl-CoA ligase; ERG13: 2-hydroxy-3-methylglutaryl-CoA synthase (HMG-CoA synthase); tHMGR: truncated 2-hydroxy-3-methylglutaryl-CoA reductase; ERG12: mevalonate kinase; ERG8: phosphomevalonate kinase; MVD1: diphosphomevalonate decarboxylase; IDI1: isopentenyl pyrophosphate–dimethylallyl diphosphate isomerase; ERG20: farnesyl diphosphate synthase, ERG9: squalene synthase, ADS: amorphadiene synthase, CYP71AV1: amorpha-4,11-diene monooxygenase, CPR: cytochrome-P450 reductase.

isomeroreductase (20–23). MEP cytidylytransferase catalyzes the formation of 4-diphosphocytidyl-2-C-methyl-D-erythritol (CDP-ME) from MEP (24–28). The enzyme 2-C-methyl-D-erythritol 2,4-cyclodiphosphate (MECDP) synthase, encoded by *ispF* in *Escherichia coli* was next found to convert CDP-ME into MECDP (24, 29–36). In contrast, however, to the MVA pathway, DMAPP is not formed from IPP; instead the gene product of *ispG* first converts MECDP into 1-hydroxy-2-methyl-2-(E)-butenyl 4-diphosphate (HMBPP), which then is converted by the gene product of *ispH* into both DMAPP and IPP (37–40).

Several terpenoids, in particular the ones that act as primary metabolites in plants, can be extracted in relatively large quantities from their natural hosts. However, secondary terpenoids, especially the ones finding pharmaceutical application, are usually available in small quantities in nature. In addition,

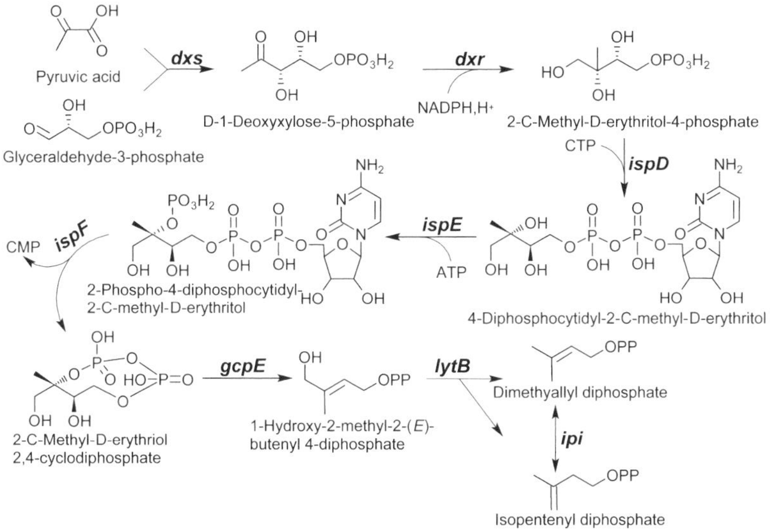

Figure 2. The *E. coli* MEP pathway for the synthesis of dimethyallyl diphosphate and isopentenyl diphosphate. The genes encoding for enzymes, indicated in bold, are: *dxs*: D-1-deoxyxylose-5-phosphate synthase, *dxr*: D-1-deoxyxylose-5-phosphate isomeroreductase, *ispD*: 2-*C*-methyl-D-erythritol-4-phosphate cytidylyltransferase, *ispE*: 4-diphosphocytidyl-2-*C*-methyl-D-erythritol kinase, *ispF*: 2-*C*-methyl-D-erythriol 2,4-cyclodiphosphate synthase, *gcpE*: 2-*C*-methyl-D-erythriol 2,4-cyclodiphosphate reductase, *lytB*: 1-hydroxy-2-methyl-2-(*E*)-butenyl 4-diphosphate reductase.

their extraction from plants depends heavily on the source plant, location, and season of harvest, and it is thus especially adversely affected by unpredictable environmental circumstances. Ever since the first mass cultivations of plant cells were achieved in the 1950s, plant cell cultures have been considered as a promising tool in the production of terpenoids (41). They have been explored for the production of various nonterpenoids, including shikonin (42) (the first secondary metabolite produced commercially), anthocyanins from *Vitis* sp., *Euphorbia milli* and *Perilla frutescens*, berberine from *Coptis japonica* and *Thalictrum minor*, rosmarinic acid from *Coleus blumei*, and hypericins from *Hypericum perforatum* (43). Because of the fact that some secondary metabolites are usually produced by specialized cells and/or at distinct developmental stages (44), hairy root cultures have also been established for the efficient production of various high-value secondary metabolites, such as alkaloids (45).

A. PLANT CELL CULTURES

In the case of terpenoids, the most successful example of employing plant cell cultures for their biotechnological synthesis is the *Taxus* cell culture production of paclitaxel, a diterpenoid commonly referred to as Taxol (46–52). It was discovered in the early 1970s after extensive characterization of extracts from northwestern United States in an effort to screen for medicinal phytochemicals. This anticancer drug is commonly used either alone or in combination with other chemotherapeutic agents (53). Similar to many other terpenoids, Taxol and its precursors are still derived from plants. Yet extraction from *Taxus brevifolia* is highly inefficient, yielding only 1 mg of taxadiene (a Taxol chemical precursor) from 750 kg of dry Pacific yew bark. Consequently, the high demand of this plant-derived metabolite has led to the depletion of natural resources, and *T. brevifolia* is now an endangered species. Today, some Taxol precursor metabolites can be chemically synthesized. However, these processes remain cumbersome and sometimes require as many as 25 steps (54, 55). As an alternative to the current production methods, plant cell culture synthesis has also been explored. In addition to Taxol, plant cell cultures have been extensively used for the production of a variety of other diterpenoids, commonly referred to as taxoids. For example, a highly efficient bioprocess strategy for the production of the taxoid taxuyunnanine C with *Taxus chinensis* cell cultures was recently reported (56–58) as well as the taxane derivatives 13-dehydroxybaccatin II and 10-deacetylpaclitaxel (59). The production of Taxol from plant cell cultures has in fact been scaled up by several companies, such as ESCAgenetic (CA), Samyang Genex (Taejon, Korea), and Phyton Biotech (Ahrensburg, Germany) (41).

Another class of terpenoids of commercial significance is pyrethrins, the most economically important natural insecticides. They encompass a set of six structurally close monoterpene esters produced by esterification of two monoterpenic acids (chrysanthemic acid and pyrethric acid) with three ketone alcohols (pyrethrolone, cinerolone, and jasmolone). They offer all the advantages of chemical insecticides, such as activity against a broad range of insects, biodegradability, as well as weak development of resistant strains and low toxicity for mammals. As such, they are acceptable as safe and environmentally innocuous alternatives to other "hard pesticides" (60). Pyrethrins are extracted from the flowers of *Chrysanthemum cinerariaefolium*, but due to increased worldwide demand, their market price has risen dramatically (61). As such, their general use in agriculture was limited because of the production costs. In the mid-1980s, the annual pyrethrin U.S. market value was estimated at $20 million with a wholesale price of $300/kg (44). The current market size is estimated at $400 million a year with a wholesale price of $413/kg (61). Based on these facts, biotechnological processes have been explored as alternative approaches to plant extraction. In that respect, pyrethrin production has been reported from callus cultures of *Tagetes erecta* (61–63) and cell suspension cultures from *Chrysanthemum coccinum* (61). In all cases, the production levels appear to be low and a number of problems require solution before such processes by plant cell/tissue cultures can become commercially profitable.

One terpenoid molecule of immense importance is artemisinin, a sesquiterpene lactone containing an endoperoxide bridge. Obtained from the Chinese plant *Artemisia annua*, it has become increasingly popular as an effective and safe alternative therapy against malaria (64). This is critical since malaria-causing *Plasmodium* strains have begun to develop resistance to traditional antimalarial compounds, such as chloroquine, cycloguanile, and sulfadoxin, whereas artemisinin remains an effective treatment option (65, 66). Its current commercial sources are field-grown leaves and flowering tops of *A. annua* (67–71). Current extraction methods for artemisinin are inefficient and result in inadequate production levels that cannot accommodate the growing global demand for inexpensive antimalarial drugs. As a result the synthetic preparation of this terpenoid has also been explored. The total synthesis of artemisinin was first described by Schmid and Hofheinz in 13 synthetic steps and a total yield of 5% (72). Ravindranathan et al. described a stereoselective synthesis of artemisinin from (+)-isolemonene (73) while Avery et al. reported a 10-step synthetic route with (R)-(+)-pulegone as the starting compound (74). Another alternative is the synthesis of artemisinin from precursor metabolites, such as artemisinic acid (75–78), the most abundant sesquiterpene in *A. annua* (78–80). In addition, chemical synthesis has also developed several artemisinin analog with potential antimalarial properties (81–83).

The general low yields obtained from organic synthesis methodologies and the complexity of the steps involved have encouraged the development of alternative biotechnological methods using in vitro plant cell and tissue cultures. In general, production of artemisinin requires a certain degree of differentiation, such as the formation of shoots and hairy roots (84, 85). Using such cultures, several attempts have been made in order to enhance production, usually by changing culture and environmental conditions, such as total nitrogen, types of carbon and nitrogen sources, presence of plant hormones, light irradiation, and temperature (85–92). For further production enhancement, genetic engineering of *A. annua* has also been explored (93–95). For example, a complementary deoxyribonucleic acid (cDNA) encoding cotton farnesyl pyrophosphate synthase placed under a CaMV 35S promoter was transferred into *A. annua* via *Agrobacterium tumefaciens* strain LBA 4404. In the transgenic plants, the concentration of artemisinin was approximately 8–10 mg/g dry cell weight (DCW), which was two- to threefold higher than that in the control (96).

B. MICROBIAL SYNTHESIS

To date, the engineering of microbial platform for terpenoid biosynthesis has largely focused on the biosynthesis of artemisinin (Figure 1B) and Taxol (Figure 3) since the demand of these important drugs exceeds production capabilities (97). In an effort to increase artemisinin availability, *E. coli* was engineered to synthesize an artemisinin precursor, amorphadiene (98). In this study, MVA kinase, phosphomevalonate kinase, and MVA pyrophosphate decarboxylase from *Saccharomices cerevisiae* together with IPP isomerase and farnesyl pyrophosphate synthase from *E. coli* were first sewn together under the control of the *lac* promoter to form a synthetic operon cloned onto a coreplicable plasmid. The synthetic operon was transformed and coexpressed with a codon-modified amorphadiene synthase (98). This resulted in a recombinant strain capable of producing up to 24 mg/L of amorphadiene. At first, containment of amorphadiene was challenging since this compound is volatile. To rectify this issue, a two-phase partitioning bioreactor (TPPB) strategy was implemented that resulted in the separation of the volatile product, amorphadiene, from the fermentation broth. With this strategy, approximately 0.5 g/L of amorphadiene could be collected (99).

Yeast has also been engineered as a production platform for artemisinin biosynthesis (100). To increase artemisinin synthesis, several genes of FPP were overexpressed. Specifically, the expression of *upc2-1*, the gene encoding for the sterol biosynthetic gene (*ERG13, ERG12, ERG8*) activator was amplified together with a truncated, soluble form of 3-hydroxy-3-methylglutaryl-coenzyme A reductase (*tHMGR*). In parallel, the yeast gene *ERG9*, which encodes for squalene synthase, was downregulated by replacing the native promoter sequence by a methione-repressible promoter. The repression of *ERG9* was performed in order to reduce carbon flux channeling

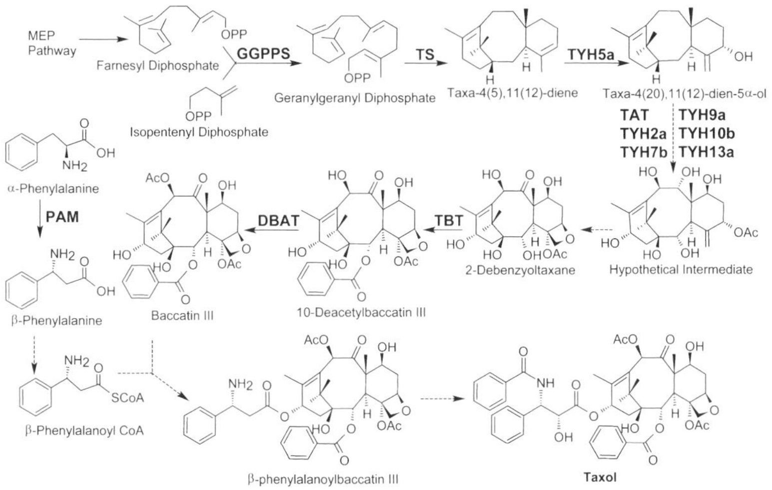

Figure 3.   Terpenoid pathway for the synthesis of Taxol. The enzymes, indicated in bold, are: GGPPS: Geranylgeranyl diphosphate synthase, TS: taxadiene synthase, TYH5a: taxadiene 5α-hydroxylase, TYH5a: taxadiene 5α-hydroxylase, TAT: taxa-4(20), 11(12)-dien-5a-ol-*O*-acetyltransferase, TYH2a: taxane 2α-hydroxylase, TYH7b: taxane 7β-hydroxylase, TYH9a: taxane 9α-hydroxylase, TYH10b: taxane 10β-hydroxylase, TYH13a: taxane 13α-hydroxylase, TBT: taxane 2a-*O*-benzoyltransferase, DBAT: 10-deacetyl baccatin III-*O*-acetyltransferase, PAM: phenylalanine aminomutase. Dashed arrows indicate several as-yet-undefined steps.

toward sterol biosynthesis. While the overexpression of *ERG20* only had marginal effect toward production improvement, the heterologous expression of amorphadiene synthase and the strategies to improve the FPP pool resulted in amorphadiene production up to 153 mg/L. Unlike *E. coli*, as a eukaryote, yeast can naturally support the functionality of membrane-bound cytochrome-P450 enzymes. To allow the synthesis of artemisinic acid from artemisinin, the P450 enzyme CYP71AV1 from *A. annua* was expressed in the artemisinin producer yeast strain together with *A. annua* cytochrome-P450 reductase (CPR). This strategy yielded the production of artemisinic acid at the level of 100 mg/L (100).

As previously mentioned, Taxol and its precursors are still derived from plants; however, extraction methods from *T. brevifolia* are highly inefficient, yielding only 1 mg of taxadiene (a Taxol chemical precursor) from 750 kg of dry bark. As an alternative to the current production methods, microbial synthesis of taxadiene has also been explored. Taxol biosynthesis starts from the isomerization of IPP to form DMAPP by IPP isomerase. In the second

step, geranylgeranyl diphosphate (GGPP) is formed from the condensation
of three molecules of IPP with one molecule of DMAPP by the enzyme
GGPP synthase. Finally, taxadiene synthase catalyzes the cyclization of
GGPP to form taxadiene (101). Currently, the numerous remaining biosynthetic steps toward Taxol formation are not fully elucidated. The first attempt
to engineer taxadiene biosynthesis in *E. coli* was performed through the
overexpression of IPP isomerase, GGPP synthase, and taxadiene synthase. In
parallel, in order to increase IPP availability, the expression of DXP synthase
was also amplified (102). These heterologous genes were cloned separately
into multiple coreplicable expression plasmids where the expression level of
each gene was regulated by the strong T7 phage promoter. Furthermore, the
functionality of the plant-derived taxadiene synthase in *E. coli* was promoted
by improving protein solubility through N-terminal truncation. Upon recombinant protein induction, the recombinant *E. coli* strain achieved production levels of up to 1.3 mg/L of taxadiene in batch fermentations, a
significant improvement over plant extraction.

  Not unlike artemisinin biosynthesis, formation of Taxol requires the use of
a membrane-bound cytochrome-P450 (103); thus attempts at biosynthesis of
downstream metabolites in the Taxol biosynthetic pathway have required the
use of recombinant *S. cerevisiae* strains. In this strategy, in addition to the
functional expression of GGPP synthase and taxadiene synthase, the three
steps following taxadiene formation are also introduced to the recombinant
yeast to produce taxadiene-5α,10β-diol monoacetate. The three additional
enzymes are taxoid 10β-hydroxylase, taxadienol 5α-*O*-acetyl transferase,
and the cytochrome-P450 taxadiene 5α-hydroxylase. Culturing the recombinant strain yielded up to 1.0 mg/L of the taxadiene intermediate, but only
trace amounts (<25 μg/L) of the diol product were produced. This indicated
the functionality of GGPP synthase and taxadiene synthase. However, this
result also highlighted the poor expression of the rest of the biosynthetic
pathway. To increase the activity of the P450 enzymes, the coordinated
overexpression of the P450 monooxygenases and P450 reductases has been
suggested (103, 104).

III.   Carotenoids

  Carotenoids are ubiquitous carbon chain molecules found in chlorophyll-
containing photosynthetic organisms as well as some bacteria and fungi.
Carotenes and xanthophylls make up the two subclasses of the more than 700

different carotenoids. These classes are distinguishable by the lack of oxygen atoms in carotenes or presence of oxygen in xanthophyll carbon chains. An increasing number of studies have shown that carotenoids may provide human health benefits; thus most of today's carotenoids are often incorporated into human diets to improve one's health. Recent clinical studies on the effects of carotenoids have shown promising roles in the treatment of cancers and heart disease and the reduction of the oxidative stresses induced by high-fat and high-cholesterol diets (105–109). Improvement of carotenoid content in fruits and vegetables has been extensively pursued through selective plant breeding and genetic manipulations (110–117). Due to the increasing demand for carotenoids, recent interest in production technologies has focused on microbial biosynthetic production of these high-value chemicals for which high-throughput screening is accommodated by their colored nature. In addition, other high-value compounds such as C_{20}-retinoids, C_{15}-phytohormones, and C_{13}-aromas are carotenoid derivatives formed by carotenoid cleavage mechanisms (for a detailed review see ref. 118).

A. CLASSIC CAROTENOIDS

Carotenoid biosynthesis in bacteria originates from the metabolism of two glycolytic precursors, pyruvate and glyceraldehyde 3-phosphate (G3P), into isopentyl diphosphate (IPP). The universal carotenoid pathway (Figure 4A) then uses multiple IPP molecules in head-to-head condensation reactions by farnesyl diphosphate synthase (IspA) and geranylgeranyl diphosphate synthase (CrtE) to form the 20-carbon chain molecule of GGPP. This is then followed by a two-step dimerization by phytoene synthase (CrtB) in which two GGPP molecules are again head-to-head condensed followed by cleavage of the diphosphate tails to synthesize phytoene. The pathway continues by phytoene desaturase (CrtI) resulting in formation of ζ-carotein, a light-activated molecule that itself is further desaturated to generate the first colored carotenoid, lycopene. Lycopene, the antioxidant commonly associated with tomatoes, is composed of 13 trans double bonds along the carbon chain causing the highly bioactive nature of the molecule. The accessibility of double bonds allows for its easy modification by a variety of enzymes to form the two classes of carotenoids. Lycopene cyclase (CrtY) is the enzyme responsible for carrying out the cyclization of each end of the chain molecule to form β- or α-carotene. Further modifications are introduced through the extension of the carbon backbone by prenyl-transferase-like enzymes, the addition of hydroxyl groups by hydroxylases and epoxidases, and even

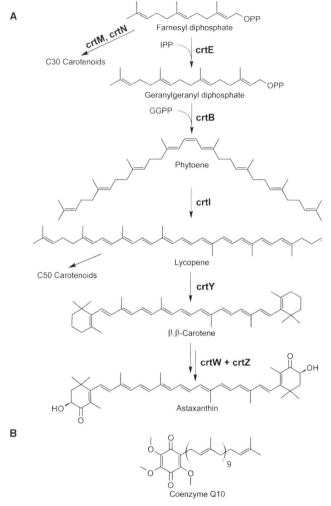

Figure 4.   (A) Carotenoid pathway for the synthesis of β-carotene and astaxanthin. (B) Molecular structure of coenzyme-Q10 (ubiquinone). IPP: Isopentyl pyrophosphate; GGPP: Geranylgeranyl diphosphate. The enzymes, indicated in bold, are: crtE: geranylgeranyl diphosphate synthase, crtB: phytoene synthase, crtI: phytoene desaturase, crtY: lycopene cyclase, crtW: β-carotene ketolase, crtZ: β-carotene hydroxylase, crtM: dehydrosqualene synthase, crtN: dehydrosqualene desaturase.

the addition of functional groups by methyl and glycosyl transferases, which are often encoded by genes within the same gene cluster as the upper carotenoid pathway (119). Alternatively, C_{30}-carotenoids arise from condensations of two 15-carbon FPP molecules, the intermediates in GGPP synthesis catalyzed by CrtM and CrtN. In yeasts, FPP is converted into ergosterol (provitamin D2), which is the principal isoprenoid molecule as it is an essential part of the yeast membrane. Thus, the major challenge in enhancing carotenoid production in yeast is the redirection of metabolic flux away from the steroid and toward carotenoids—challenging task since ergosterol is critical to cell growth.

Elucidation of carotenoid biosynthetic pathways has allowed the engineering of noncarotenogenic industrial strains such as baker's yeast *S. cerevisiae* and *E. coli* for carotenoid production. This is typically achieved by simultaneous functional expression of the carotenoid biosynthetic enzymes that have been identified and isolated from the known carotenogenic species (120–130). Biosynthesis of lycopene was achieved through the expression of CrtB, CrtE, and CrtI from the bacterium *Erwinia uredovora* under the control of a *S. cerevisiae* promoter. In order to achieve the functionality of the bacterial enzymes in yeast, the translation codons of the bacterial genes were modified without changing the amino acid sequences. The engineered recombinant produced lycopene up to 113 µg/g dry weight where a similarly engineered *S. cerevisiae* harboring an additional plasmid containing lycopene β-cyclase from *E. uredovora* (CrtY) resulted in production of up to 103 µg/g dry weight of β-carotene (131).

Recently, studies have focused on demonstrating the feasibility of elevating lycopene biosynthesis from an engineered *E. coli* strain. Overexpressing 1-deoxy-xylulose 5-phosphate synthase (DXPS), FPPS, isopentenyl diphosphate isomerase (Idi), and the *crtEBI* operon are aimed at increasing the conversion of the glycolytic metabolites pyruvate and glucose-3-phosphate into IPP. One such study used these overexpressions and constraint-based modeling through flux balance analysis and minimization of metabolic adjustment (132, 133) to optimize lycopene biosynthesis in the recombinant *E. coli* strain. Modeling was used to search for single gene deletion targets within the entire genome that resulted in elevated production levels. Additionally, gene deletions by transposon-based mutagenesis (134) identified an additional three gene deletions that improved lycopene biosynthesis (135). Identified were the *rssB*, a gene controlling macromolecule degradation, and two hypothetical proteins, YifP and YjiD. Combining the single transposon-based mutants with gene deletions predicted computationally into combinatorial deletion strain produced lycopene over 10 mg/g DCW (136). Similarly, shot-gun library clones of a lycopene producing *E. coli* strain using colorimetric screening identified six colonies with improved lycopene yields (137). Four genes (*dxs, appY, crtI,* and *rpoS*) were identified among the clones to be involved in the production enhancement. Coexpression on cloned vectors for these genes synergistically enhanced production of lycopene from 0.6 mg/g DCW up to 4.7 mg/g DCW with *appY* and *dxs* overexpressed (137). The role of *dxs* and *rpoS* in carotenoid accumulation was found in an earlier study where elevated *rpoS* transcription directly correlated to increased carotenoid production (138). In fact, a study of the green algae *Haematococcus pluvialis* showed accumulation occurred in the developing stages due to ketocarotenoid accumulation (139).

Another attractive carotenoid for commercial use is astaxanthin, an abundant pigment found in marine animals such as salmon and crustaceans, which has biological activity as an antioxidant, immune system modulator, and, as suggested by some studies, anticarcinogenic and antitumor activities (140–142). Astaxanthin is synthesized from β-carotene through a variety of ketocarotenoid intermediates using only β-carotene ketolase (CrtW) to add the keto groups and β-carotene hydroxylase (CrtZ) to hydroxyl groups to the ringed structure. The isolation and characterization of these two genes from *Paracoccus haeundaensis* led to their functional expression by first cloning the whole gene cluster encoding proteins CrtWZYIBE into pCR-TOPO and then transforming into *E. coli* BL21 to enable high protein expression levels. The recombinant strain produced 400 μg/g DCW of astaxanthin, accounting for almost 70% of the total carotenoid levels in the strain (143, 144). The isolation of a new CrtW from an environmentally isolated *Sphingomonas* sp. DC18, along with localized random mutagenesis, enabled the production of astaxanthin to reach 90% of the total carotenoids by a recombinant *E. coli* strain (145).

In addition to enabling gene expression, proper regulation of the carotenoids metabolically engineered pathway is important. It has been shown that the overexpression of phosphoenolpyruvate synthase to direct excess pyruvate into G3P and other precursor balancing are critical for enhanced carotenoid production in recombinant strains (146). Oxygen availability is another factor that dramatically affects astaxanthin biosynthesis in the methanotrophic bacterium *Methylomonas* sp. 16a. By coexpressing proteins CrtW and CrtZ with bacterial hemoglobin to increase oxygen, availability within the cell levels of astaxanthin reached up to 60% of total carotenoid content compared to only 10% without the hemoglobin. They also showed low production under low-dissolved-oxygen (DO) levels with the ketocarotenoid intermediates at high-levels, but under moderate DO levels, carotenoid production was shifted toward astaxanthin biosynthesis (147). Native producers of astaxanthin have also been engineered in attempts to create viable production platforms with productivity levels as high as 5.6 mg/L per day in continuous cultures (148). Promoter replacement has also been explored for enhanced carotenoid production in *E. coli* by swapping the native promoters of carotenoid genes with the stronger bacteriophage T5 promoter. After introduction of the T5 promoter before *dxs*, *ispD*, *ispF*, *idi*, and *ispB* genes, beta-carotene production increased to 6 mg/g DCW (149).

B. UNNATURAL CAROTENOIDS

Extension of the carotenoid family has been investigated by engineering the natural carotenoid pathway to generate novel compounds, most notably novel (hydroxylated) acyclic and cyclic carotenoids, which may have additional nutritional and dietary benefits. Carotenoid cyclases are the primary targets for introducing diversity into the carbon skeletons (150); however, engineering unique desaturases to form extremely long backbones has also been explored (151). Eight different hydroxylated carotenoids were produced in *E. coli* transformed with a combination of up to three compatible plasmids containing several carotenogenic genes from *E. uredovora* and two *Rhodobacter* species. Recombinant strains yielded concentrations ranging from 41 to 176 µg/g DCW including acyclic molecules 1-hydroxyneurosporene, 1-hydroxylycopene, 1,1'-dihydroxylycopene, and demethylsperoidene as well as the cyclic molecules 3-hydroxy-β-zeacarotene, 7,8-dihydrozeaxanthin, 3- or 3'-7,8-dihydro-β-carotene, and 1'-hydroxy-γ-carotene; most are only found in trace amounts from natural sources. In general, carotenoid concentrations were found to be higher at lower temperature, an inverse relationship of carotenoid formation observed before in *E. coli* (152). The negative association between growth and carotenoid production can be attributed to both a slower gene expression yielding more active enzymes and better conditions for carotenoid precursor availability when the entire metabolism is slower (152).

Use of gene shuffling with genes encoding phytoene desaturases derived from *Erwinia herbicola* and *E. uredovora* resulted in a library for introduction into *E. coli* harboring wild-type GGPP synthase and phytoene synthase (153). Subsequent selection of clones exhibiting alternative carotenoid colorations, particularly clones with yellow coloration (I25) and pink coloration (I14), were selected. Sequence analysis of the mutated CrtE isolated from I14 and I25 clones revealed two different amino acid mutations and, in the case of I14, the replacement of the 39 N-terminus amino acids of CrtE from *E. uredovora* with those of *E. herbicola*. Continued expansion of the novel cyclic carotenoid library was achieved by separately introducing CrtY from *E. uredova* and *E. herbicola* into the carotenoid pathway containing CrtE from clone I14. A bright yellow-orange coloration was produced using the wild-type desaturase in the recombinant *E. coli*; however, replacement of the wild type by the mutated desaturase from I14 resulted in bright yellow coloration. Similarly, a library of lycopene cyclases was also created by shuffling of different *crtY* genes and screened to yield 25 colonies out of 4500 with different colorations. A reaction product isolated from a bright red clone was identified to be torulene, a compound not native to the recombinant metabolic pathway introduced (153). A similar approach has recently been presented where carotenoid production alterations from *E. coli* by random chromosomal mutations were used to produce novel carotenoids (154).

C. UBIQUINONE

While not a true carotenoid, coenzyme Q10 (CoQ10), also referred to as ubiquinone, uses the same isoprene units to make its tail that are used for the hydrocarbon chain of traditional carotenoids (Figure 4B). CoQ10 is widely regarded as one of the most important lipophilic antioxidants that can prevent the generation of free radicals as well as oxidative modifications of proteins, lipids, and DNA. Many human pathological conditions are associated with reduced levels of CoQ10, including cardiac disorders, neurodegenerative diseases, and cancer, all of which are usually treated with dietary CoQ10 supplements. To satisfy demand for this important fine chemical, CoQ10 has been synthesized by conventional chemical synthesis (155), by semi chemical synthesis (156), and more recently from native and recombinant microbial strains (157, 158).

Early studies of bacterial production led to the identification of three efficient CoQ10-producing strains from the species *A. tumefaciens*, *Rhodobacter sphaeroides*, and *Paracoccus denitrificans* and the discovery of two mutated overproducer strains (158). After an initial screening for high natural production, additional random mutations were then introduced in two *A. tumefaciens* strains by chemical treatment using *N*-methyl-*N'*-nitro-*N*-nitrosoguanidine (NTG). Following 90-h fermentations, CoQ10 production in mutant strains increased from 50 mg/L in the wild-type *A. tumefaciens* up to 110 mg/L in the mutant strains. The higher concentrations were attributed to the introduced genetic alterations, although mutation sites were never identified (158). Developments in recombinant DNA (rDNA) technology helped to further expand biosynthesis of CoQ10 through multiple gene insertions. Park and colleagues performed batch and fed-batch fermentations of *E. coli* BL21 strains episomally expressing decaprenyl diphosphate synthase (encoded by *ddsA*) from *Gluconobacter suboxydans* (157). The *ddsA* gene was introduced to the cells by two different coreplicable plasmids: a high-copy-number plasmid (pUC19) and a low-copy-number plasmid (pACYC184). The low-copy-number pACDdsA strain consistently outperformed the high-copy-number pYCDdsA strain to achieve CoQ10 production levels of 0.97 mg/L in batch cultures and a final concentration of approximately 25.5 mg/L under fed-batch conditions. Productions levels, while still low, illustrate the limited effort needed to achieve competitive titers through microbial production of this complex isoprenoid.

IV. Polyketides

Polyketides are a large class of natural products with a wide range of chemical structures that impart a range of physical properties suited for a variety of applications, particularly in pharmaceuticals (Figure 5). These druglike molecules constitute one of the biggest sources for natural product–based therapeutics since many of them act as potent antibiotics (erythromycins) and antitumor agents (epothilones) (159–161). The polyketide market is a $20 billion market, making them second only to the penicillins in terms of their importance as naturally derived medicines (162). Unlike flavonoids and carotenoids, there is no one route of synthesis for these complex molecules. Instead, polyketides are distinguishable by which polyketide synthases (PKSs) are used to perform the proper chemical manipulations using simple small molecules including acetyl-CoA, malonyl-CoA, methylmalonyl-CoA, and propionyl-CoA.

The main challenge for total chemical synthesis of many polyketides involves stereoselective C−C bond formations. The first totally chemical route for synthesis of erythromycin A was performed using lithium enolates and a stereoselective aldol reaction to form the C2−C3 bond (163). At the same time, using similar aldol reactions but with chiral boron enolates, 6-deoxyerythronolide B was formed (164). Today, a variety of enolates

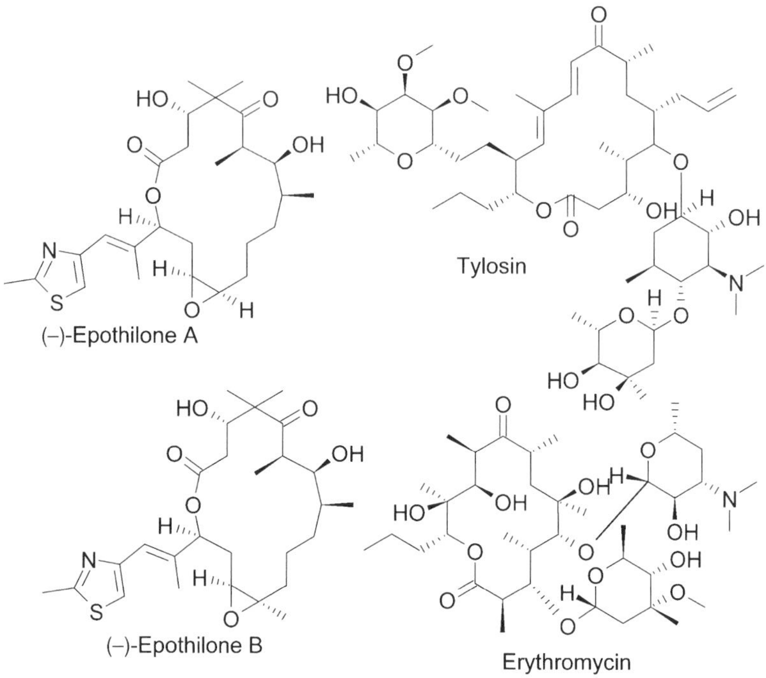

Figure 5. Molecular structures of the polyketides epothilone A, epothilone B, tylosin, and erythromycin.

have been used to achieve stereoselectivity and conformation (165–168). However, before recent developments in C–C bond-formation processes, particularly the aldol reaction (for review see ref. 169), it was noted that the polyketide representative erythromycin A seemed "quite hopelessly complex, particularly in view of its plethora of asymmetric centers" (170).

Similar to bacterial polyketides, plant polyketide molecules have also found many important medical and industrial applications. For example, benzalacetote is the precursor of many functional chemicals such as the anti-inflammatory lindleyin found in rhubarb, gingerol and curcumin found in ginger plants, and raspberry ketone, the chemical responsible for raspberry's characteristic aroma. Plant polyketide biosynthesis is mediated by type III polyketide synthases (PKSs) (extensively reviewed in ref. 171). Interestingly, the adaptability of plant PKSs allows for their engineering to generate novel functions and catalytic activities and specificities. For example,

bioinformatic studies have highlighted critical differences within the catalytic region of plant PKSs. Using a secondary-structure alignment
of chalcone synthase (CHS), stilbene synthase (STS), 2-pyrone synthase
(2-PS), and acridone synthase (ACS), it was shown that the conserved amino
acid residue Phe-215, thought to be a crucial integral of the catalytic activities
of CHS, is not present in the benzalacetone synthase (BAS) gene. By
replacing the amino acid Leu-215, together with its adjacent Ile-214 with
Phe-215 and Leu-214, respectively, it was shown to confer CHS activities to
an enzyme naturally exhibiting BAS activity. Mutations of BAS resulted in
chalcone-forming properties, in which naringenin chalcone, along with other
by-products, were generated from incubation with appropriate substrates
(172). Through this example, it can be shown that these bioinformatic
endeavors are vital to biocatalyst engineering. In addition to secondary-
structure analysis, such mutational study is often supported by structural data.

A. EPOTHILONES

Epothilones have been the target of many different organic syntheses
approaches with the first published total synthesis of epothilones in 1996
(173) with other epothilones published since (174, 175). All total synthesis
approaches first construct the key building blocks aldehyde, glycidol, and
ketoacid, which are then coupled to olefin metathesis precursor via an aldol
reaction followed by esterification. Closing the bis terminal olefin using a
Grubbs catalyst then results in compounds with distinct stereocenters
having cis- and trans-macrocyclic isomers. Recently, synthesis of a number
of epothilone analogs has occurred using both chemical synthesis to generate
alternative thiazole appendages (176) and open chains incorporating the
C1–C8 fragment and aromatic rings (177) and recombinant organisms using
different starting units (178–180). Due to the size of the PKS, functional
expression in *E. coli* is difficult.

Despite the recent successes in chemical synthesis, industrial development of polyketides via traditional chemical means often suffers from a
number of complications that lead to excess costs, including rapid polymerization resulting in the formation of unwanted products and troublesome site-
directed synthesis. Biosynthesis of many potent polyketides can overcome
the stumbling blocks in chemical synthesis; however, the cell cultures used
tend to be slow growing and/or difficult to culture, making it inconvenient
for efficient large-scale production. Thus, several industrial microorganisms
have been engineered as alternative production platforms to produce

polyketide molecules. As such, recombinant strains must fully express the PKSs responsible for the proper folding and posttranslational modifications needed to create the complex molecules. Simple overexpressions of encoding genes is usually not enough and further alterations, typically using metabolic and protein engineering techniques, are used to modify enzymes or transform plasmids. In addition, since the small molecules used as starting materials are readily used throughout the cell, increasing the availability of the precursors is also a priority, especially in dense cell cultures. Further complicating polyketide production is that these two issues need to be resolved at the same time for efficient production (181). These issues are being addressed in a new generation of microbial strains for polyketide production.

Epothilones are complex ring molecules with a methylthiazole group connected to the macrocycle. Biosynthesis of epothilone B, the natural form, is performed by a type I PKS that generates the backbone while the thiazole ring is included via cysteine incorporation by a nonribosomal peptide synthetase (nRPS). Epothilone B biosynthesis begins with a methylated carboxythiazole starting molecule, which is formed through translational coupling of the *epoA*-encoded EPOSA of the PKS with the *epoP*-encoded EPOSP of nRPS. Both of these modules have multiple functional domains which are critical in forming the thiazoline ring by intramolecular cyclodehydration. To do this, EPOSP activates a cysteine which then binds to an aminoacyl-*S*-peptidyl carrier protein, at which point EPOSA shuttles an acetate unit onto the EPOSP complex to initiate thiazoline ring formation. Once the ring is formed it is transferred to the PKS for subsequent elongation and functionalization using the rest of the biosynthetic cluster encoded by genes *epoB*, *epoC*, *epoD*, *epoE*, and *epoF*. A final P450 epoxidase encoded by *epoK* is responsible for the epoxidation of epothilones C and D (182).

Biosynthetic routes for production of epothilones have been discovered through the use of ^{13}C labeling experiments not only to trace origins but also to identify the enzymatic mechanism for the pathway (183). Interestingly, it has been shown that the PKS end products epothilones C and D are in competition for binding to the active site of the P450 epoxidase for further functionalization to form the epothilones A and B, a process once thought to occur using different PKSs (184). Due to the slow doubling time of *Sorangium cellulosum* (16 h), engineering epothilone production in faster growing recombinant hosts is a commercially viable endeavor. The gene cluster encoding the PKS and the P450 from the myxobacterium *S. cellulosum* was functionally expressed in *Streptomyces coelicolor* by concomitant expression, enabling 10 times faster epothilone production (185). *Myxococcus xanthus* strains have also been used for the biosynthesis of epothilones up to 17 μg/L, where the genes were inserted through homologous recombination events. Compared to epothilones produced in *S. cellulosum*, levels of epothilone B are about 100-fold lower in *M. xanthus* under similar growth conditions. Interestingly, the ratio of epothilone A to epothilone B dramatically changed to approximately 1 : 10 in *M. xanthus* as compared to a roughly 2 : 1 ratio in *S. cellulosum* (186). In a parallel study, an optimized cell culture of *M. xanthus* was able to produce an average production titer of 23 mg/L of epothilone D in dense cell cultures, although it required the addition of an end-product stabilizer (Amberlite® XAD-16) at 20 g/L concentrations (187). Recently however, the use of lowered expression temperature, chaperone coexpression, use of the P_{BAD} promoter, and additional engineering by gene fusions to express the *epoD* cluster epothilone C and D were identified by mass spectrometry (188).

B.   ERYTHROMYCINS

Erythromycin is a potent polyketide antibiotic synthesized by the soil bacterium *Saccharopolyspora erythrea*. The macrolytic antibiotic core is synthesized by large modular type II PKSs in which the core molecule 6-deoxyerythronolide B (6dEB) is formed from one propionyl-CoA unit and subsequent elongation of six (2*S*)-methylmalonyl-CoA by the enzyme deoxyerythronolide B synthase (DEBS). DEBS is a three-subunit enzyme ($\alpha_2\beta_2\gamma_2$) comprising two sets of 28 distinct active sites, 7 of which are modified after translation by pantetheinylation (189). As noted earlier, this complex chemical structure hampers the total chemical synthesis of this antibiotic. Expression of DEBS genes has been achieved in recombinant *S. coelicolor* (190). However, since the scalability of *Streptomyces* could still be a rate-limiting step, the engineering of an easy-to-culture microbe such as *E. coli* was also pursued.

To engineer polyketide synthesis, the three subunits of DEBS were cloned individually into the *E. coli* expression vector pET21c. In order to facilitate pantetheinylation of the recombinant DEBS and synthesis of propionyl-CoA, a phosphopantetheinyl transferase gene (*sfp*) and propionyl-CoA synthase (*prpE*) were also inserted into *E. coli* by integration into the *prpRBCD* operon within the *E. coli* genome (181). Disruption of the *prp* operon, which is responsible for propionate metabolism, was intended to allow optimum conversion of exogenously supplemented propionate into propionyl-CoA by the *prpE* gene product. Furthermore, the two-subunit propionyl-CoA carboxylase (*pcc*) and the biotin ligase carrier protein (*birA*) genes were also introduced into the recombinant strain mediated by a coreplicable plasmid. Introduction of the carboxylase gene allowed the conversion of propionate into (2S)-methylmalonyl-CoA, which served as an extender unit for the recombinant DEBS. Fermentation of the highly engineered recombinant strain in propionate-supplemented media yielded 0.1 mmol of 6dEB per gram of cellular protein per day, which is superior to wild-type *S. erythraea* and comparable to a modified strain used for industrial 6dEB production (181).

Novel polyketides, especially erythromycins, have been synthesized in *E. coli* using modified PKSs with modified promoters or where cyclases and other functional domains are repositioned to redirect polyketide synthesis and generate novel molecules (191). In an earlier study, two amino acid substitutions were introduced in the putative NAD(P)H binding motif located in the proposed enoyl reductase domain of DEBS. The resulting unnatural product was identified as Δ6,7-anhydroerythromycin C, a result caused by the lack of enoyl reduction that typically occurs during the synthesis of the macrolactone (192). Another study relocated the chain-terminating cyclase domain of DEBS isolated from *S. erythraea* to the carboxyl-terminus of DEBS1, the multienzyme that catalyzes the first two rounds of the polyketide chain extension. As a result of the repositioning, the formation of the predicted triketide lactone was greatly accelerated compared to an inactive control (193). In a similar study, one of the functional units of erythromycin PKS was first replaced by removing a methylmalonate-specific acyltransferase domain, responsible for formation of the methyl side chain at C_6, with an ethylmalonyl-specific acyltransferase used for niddamycin biosynthesis. The engineered protein domain enabled the production of an erythromycin-like product. Following the expression of a gene encoding for crotonyl-CoA reductase, the recombinant strain was able to produce the desired 6-ethylerythromycin product (194). Combinatorial biosynthesis has also been used where a genetic block was first introduced in the initial condensation step formation of 6dEB, which was then followed by exogenous addition of designed synthetic molecules to small-scale cultures. This process resulted in highly selective milligram production of multiple unnatural polyketides, including aromatic and ring-expanded variants of 6dEB (195).

C.   TYLOSINS

The commercial production of the complex polyketide antibiotic tylosin was recently accomplished by adapting DNA shuffling to perform an entire genome shuffling of the bacterium *Streptomyces fradiae* (196). In order to generate a new tylosin overproducer, the wild-type *S. fradiae* genome was subjected to one round of chromosomal random mutagenesis by exposure to nitrosoguanidine mutagen. Upon screening of 22,000 individual mutants, 11 strains producing more tylosin than the wild-type were isolated. Then, to generate a genome-shuffled library, protoplasts of the 11 strains were mixed in equal proportion and recursively fused, after which 1000 new clones were screened from the first round of genome shuffling with 7 identified as superior strains. These were then used as the parental strains for a similar final round of shuffling and screening. Analysis of two from the final seven overproducer strains isolated in the last round showed tylosin titers ninefold higher compared to the wild-type *S. fradiae*. It is noteworthy that the development of a similar overproducer strain using various mutagens took place over 20 years, requiring one million assays while application of the genome shuffling method achieved the creation of an overproducer strain in the course of 1 year with only 24,000 assays (196).

Traditional mutagenesis via UV irradiation and nitrosoguanidine were used to increase antibiotic activity of an *S. fradiae* strain in an effort in increase tylosin production. Variants formed produced tylosin up to 28.3% over the original strain where the most active were formed by a dual treatment of both mutational methods (197). A more recent study heterologously expressed tylosin PKS in *Streptomyces venezuelae* strain in which the pikromycin PKS gene cluster was deleted. Using low-copy-number plasmids and optimized culture media, the engineered strain produced 1.4 g/L in four-day cultures, a period considerably shorter than other *Streptomyces* hosts (198). This ability for high expression of the *S. venezuelae* strain may be due to regulation of the TylR activator protein, a member of the SARP family. It has been shown that *tylS* and *tylU* disruptions in *S. fradiae* reduce expression of *tylR* and contribute to lower levels of tylosin production (199, 200).

V.   Aromatic Derivatives

A.   SHIKIMATES

Shikimic acid and quinic acid are intermediate metabolites of aromatic amino acids. These compounds have become the essential chiral starting materials of several important drugs, such as neuraminidase inhibitors for influenza treatment, the antitumor agent esperamicin-A, and the immuno-suppressant FK506. Today, even though quinic acid is readily available from

Cinchona bark, the natural supply of shikimic acid still relies on extraction from the fruit of *Illicium* plants, which is inefficient for large-scale production. Recently, the shortage of drugs needed to combat a bird-flu pandemic has motivated engineering efforts to produce these hydroaromatic metabolites from bacteria in order to increase the global drug supplies.

Shikimic and quinic acids are derived from glucose metabolism (Figure 6). Specifically, phosphoenolpyruvate (PEP) and erythrose 4-phosphate (E4P) are converted into 3-deoxy-D-arabino-heptulosonic acid 7-phosphate (DAHP) by the enzyme 3-deoxy-D-arabinoheptulosonic acid 7-phosphate synthase (AroF). DAHP is subsequently converted into 3-dehydroquinic acid (DHQ) by 3-dehydroquinate synthase (AroB). The enzyme shikimate dehydrogenase (AroE) then catalyzes the conversion of DHQ into quinic acid. The reversible reaction catalyzed by 3-dehydroquinate dehydratase (AroD) converts DHQ into 3-dehydroshikimic acid (DHS). AroE catalysis is also responsible for the conversion of DHS into shikimic acid. Shikimic acid is further converted into shikimate 3-phosphate by shikimate kinase (AroKL).

For the generation of *E. coli* strains capable of synthesizing high-level shikimate and quinic acid, the *aroB* gene was inserted into the *serA* locus

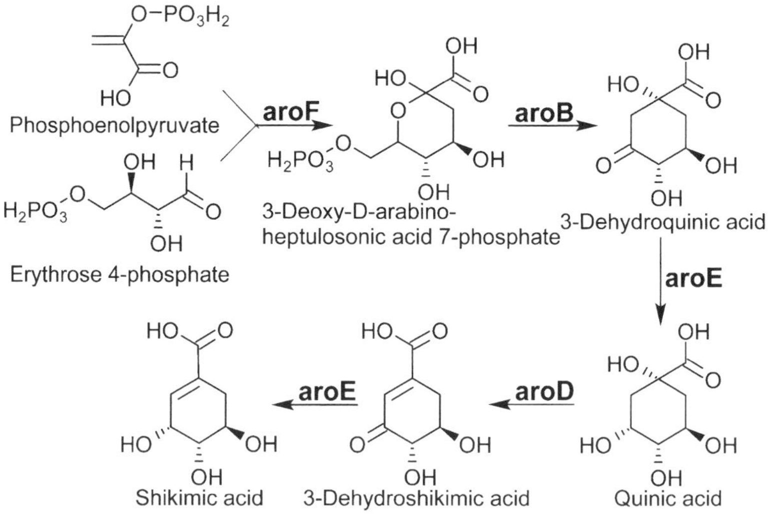

Figure 6.   Shikimic acid pathway for the synthesis of shikimic acid. The genes encoding for enzymes, indicated in bold, are: *aroF*: 3-deoxy-D-arabinoheptulosonic acid 7-phosphate synthase; *aroB*: 3-dehydroquinate synthase; *aroD*: 3-dehydroquinate dehydratase; *aroE*: shikimate dehydrogenase.

in the *E. coli* genome, while the gene *aroL* and *aroK* loci were disrupted via successive P1 phage-mediated transduction of *aroL*478::Tn10 and *aroK*17::Cm^R (201). Additionally AroF^{FBR}, a mutant AroF which is insensitive to feedback inhibition of aromatic amino acids, was overexpressed in conjunction with AroE. Cultivation of the recombinant strain resulted in the synthesis of 27.2 g/L shikimic acid, 12.6 g/L quinic acid, and 4.4 g/L DHS. In order to reduce the biosynthesis of quinic acid, AroD was also overexpressed. However, this overexpression did not result in the decline of DHQ concentration, which indicated spontaneous reversibility of shikimic acid. Overall, the high production titers indicated that the biochemical means of producing shikimic and quinic acids could be an alternative approach to the plant extraction methods. However, this biochemical production method was soon abandoned due to the development of a cost-effective synthetic method.

B. FLAVONOIDS

Flavonoids are a diverse group of plant secondary metabolites found ubiquitously in the plant kingdom. All flavonoids are related through a three-ring structure but the different subclasses, including chalcones, flavanones, flavones, flavonols, and anthocyanins, differ from each other based on the degree of bond saturation, hydroxylation, and in the case of isoflavones ring position. Growing scientific evidence demonstrates the potent health-promoting activities of many flavonoids (202, 203). However, flavonoid availability for human consumption is a concern. Even though many flavonoids are contained in edible plant products, diets in many parts of the world are low in fruits and vegetables—preventing sufficient intake of flavonoids (204). Moreover, some of the very promising flavonoids only exist in minute quantities in minor food groups such as herbs, which exacerbates the low bioavailability of these plant chemicals.

Today, some flavonoid compounds are available as nutraceutical supplements such as isoflavone containing food commodities that are popular for treating hormone-related disorders. Flavonoids are traditionally derived from plant extraction to meet general public needs. However, due to the low flavonoid concentration in planta, abundant natural resources are required for large-scale production for nutraceutical supplements. To resolve this problem, certain plants have been genetically engineered by increasing the activities of flavonoid biosynthetic enzymes. In addition, bioreactor-based systems for mass production of flavonoids in general and anthocyanins in particular have been described for a few species (205, 206), but to date economic feasibility has not been established, in part because of engineering challenges in mass cultivation of plant cultures. However, the existence of competing pathways in plants complicates the substantial increase of content of specific flavonoid compounds (207). For that reason, blocking of competing pathways had to be implemented in order to further increase flavonoid content (208). Plant cultivation also depends heavily on environmental, seasonal, and geological conditions. Therefore, consistent quality and quantity of plant resources could present a rate-limiting step to large-scale production. In the down stream processing line, flavonoid extraction and purification are also inefficient due to contamination of numerous plant small molecules and the loss of products due to processing conditions (209).

The increasing demand for flavonoid consumption drives the search for efficient large-scale production platforms. A commonly chosen production alternative of natural products is through total or partial chemical synthesis. Chemical synthesis has been reported for a number of natural flavonoids and unnatural analogs. For example, a Claisen–Schmidt condensation reaction between an acetaphenone and a benzaldehyde can lead to the formation of chalcones (210, 211), flavanones (210, 211), flavones (211, 212), and flavonols (211, 213). Stereospecific catalysts were used to achieve different chiral forms of catechins (214) and leucoanthocyanidins (215). In addition, elegant methods have been designed to make the stilbene resveratrol (216–218). In general, the total synthesis of many complicated natural products is often not feasible due to the lack of mechanistic information. When reaction mechanisms are available, the synthesis of complicated structures requires several steps. The multiple chemical steps are feasibly conducted in bench-scale experiments; however, the one-pot synthesis route is an important parameter that needs to be elucidated in order to implement efficient large-scale production. The requirement for high-energy and toxic chemicals in the process of converting reactants to products presents another disadvantage of chemical synthesis. Moreover, several chemical modifications such as glycosylation and acylation, which are commonly associated with plant products, are currently unattainable through chemical synthesis.

While many instances have demonstrated the feasibility of efficient production of plant-derived natural products through plant cell cultures, frontiers of flavonoid production have focused on heterologous synthesis using the well-characterized yeast *S. cerevisiae* and gram-negative bacterium *E. coli*. The microbial factories pose several advantages due to rapid growth, ease of cultivation, and convenient genetic manipulations. Due to these features, high-level production could be maintained. Moreover, the enzyme-based production increases product selectivity and reduces the usage of toxic chemicals while conserving energy usage. The important pharmacological and industrial properties of these phytochemicals and the limited availability of purified forms from plants have motivated the engineering of flavonoid biosynthesis in microbial hosts.

1. Flavanones

Flavanones, the direct precursors of many flavonoids, are synthesized from the amino acid phenylalanine or tyrosine (Figure 7). The enzyme phenylalanine/tyrosine ammonia lyase (PAL/TAL) converts these amino acid–building blocks into phenylpropanoic acids. Subsequently, CoA-esters are synthesized from phenylpropanoic acids by the action of propanoyl-CoA ligase. The plant polyketide synthase chalcone synthase (CHS) then catalyzes the condensation of three malonyl-CoA moieties per one CoA-ester molecule to synthesize one chalcone molecule. Chalcones are then stereospecifically isomerized into (2S)-flavanones by chalcone isomerase (CHI). *Escherichia coli* and *S. cerevisiae* were both engineered to afford the biosynthesis of plant-specific flavanones by simultaneous expression of plant biosynthetic enzymes. However, the low production titers prohibited large-scale synthesis (219, 220). Recently, extensive metabolic engineering of *E. coli* was pursued in order to enable high-level synthesis of flavanones (221). Specifically, various strategies were explored in order to amplify malonyl-CoA, the rate-limiting metabolite in *E. coli* that serves as a flavonoid building block (Figure 8). Malonyl-CoA is synthesized from acetyl-CoA by the enzyme acetyl-CoA carboxylase (ACC). The overexpression of ACC is a rational approach to increase malonyl-CoA content in *E. coli*. However, the overexpression of the *E. coli* ACC was reported to be detrimental to cell viability. In order to bypass this phenomenon, the four subunit genes *accABCD* of *Photorhabdus luminescens* ACC were heterologously expressed. Transcription of *accA* and *accD* were individually controlled under the T7 phage promoter while the transcription of *accB* and *accC* was regulated together under one T7 phage promoter sequence. By combining this strategy with the expression of *Petroselinum crispum* 4-coumaroyl-CoA ligase (4CL), *Petunia hybrida* CHS, and *Medicago sativa* CHI, flavanone synthesis reached 196 mg/L. The functionality of ACC requires biotinylation by the action of biotin ligase (BirA). BirA is known to be able to recognize ACCs from different organisms. Therefore, in the recombinant system, the endogeneous *E. coli* BirA could catalyze biotin attachment to the

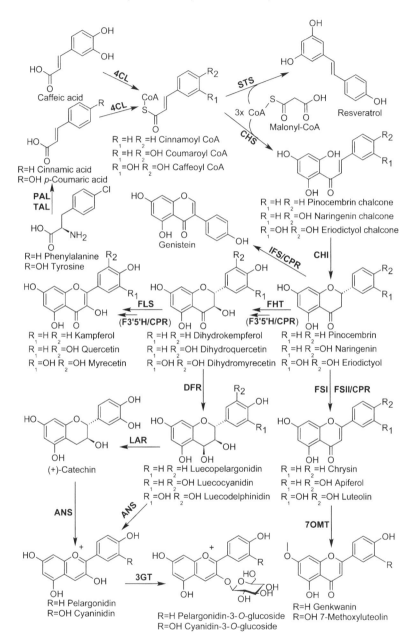

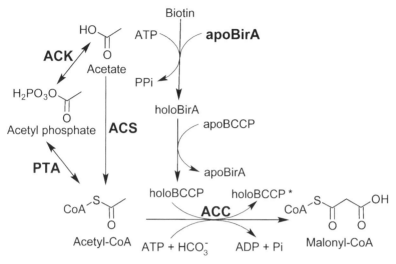

Figure 8.   Two pathways for the formation of malonyl-coenzyme A. The enzymes, indicated in bold, are: ACK: acetate kinase; PTA: phosphotransacetylase; ACS: acetyl-CoA synthase; BirA: biotin ligase; ACC: acetyl-CoA carboxylase.

P. luminescens ACC. However, in order to optimize the functionality of the overproduced ACC, *P. luminescens* BirA was also heterologously expressed. The coexpression of *P. luminescens* ACC and BirA further improved flavanone yields to 367 mg/L. Interestingly, when the *E. coli* BirA was overexpressed together with *P. luminescens* ACC, flavanone improvement decreased. The decrease in productivity could be rescued when *P. luminescens* ACC was coexpressed with a chimeric BirA containing the N-terminus from *E. coli* BirA and the C-terminus of *P. luminescens* BirA. These results suggest that the protein interaction between ACC and BirA was important for optimum malonyl-CoA synthesis.

Figure 7.   Flavonoid pathway for the synthesis of stilbenes, flavanones, flavones, flavonols, isoflavonoids, and anthocyanins. The enzymes, indicated in bold, are: PAL: phenylalanine ammonia-lyase; TAL: tyrosine ammonia-lyase; 4CL: 4-coumaroyl–CoA-ligase; STS: stilbene synthase; CHS: chalcone synthase; CHI: chalcone isomerase; FHT: flavanone 3β-hydroxylase; FLS: flavonol synthase; F3'5'H: flavonoid 3',5'-hydroxylase; FSI: flavone synthase; FSII: flavone synthase; DFR: dihydroflavonol 4-reductase; LAR: leucoanthocyanidin synthase; IFS: isoflavone synthase; CPR: cytochrome-P450 reductase; 7OMT: 7-O-methyltransferase; ANS: anthocyanidin synthase; 3GT: UDPG-flavonoid 3-O-glucosyl transferase.

The improved flavanone synthesis by the coexpression of ACC and BirA also depended on biotin supplementation. Since biotin is an expensive metabolite, further metabolic engineering of *E. coli*'s central metabolism was also explored to allow the conversion of inexpensive metabolites into high-value flavanones. In this case, the amplification of two acetate assimilation pathways in *E. coli* was explored. Acetate is a toxic fermentative metabolite. In *E. coli*, acetate is assimilated back into acetyl-CoA via two pathways. The first pathway is composed of acetate kinase (ACK) and phosphotransacetylase (PTA), and the second pathway consists of acetyl-CoA synthetase (ACS). The engineered overproduction of ACK and PTA together with ACC only resulted in a moderate increase in flavanone synthesis, presumably due to the reversibility of the ACK/PTA pathway. However, the coexpression of ACC with ACS resulted in flavanone production up to 383 mg/L along with a decrease in acetate accumulation. Exogenous supplementation of acetate at a low level further improved flavanone synthesis to 429 mg/L. The combination of these strategies resulted in the highest recombinant flavanone synthesis thus far.

2. *Isoflavones*

Their unique core chemical structure distinguishes isoflavones from the rest of the flavonoid subgroups, for which they are often referred to as "phytoestrogens," due to the resemblance to the human hormone 17β-estradiol and the ability to exert estrogen-like activities. An expansive collection of scientific publications based on epidemiological and laboratory animal evidence and mechanistic data has concluded the potential clinical importance of isoflavonoids, especially isoflavones, to promote health and prevent or delay the onset of certain chronic diseases. Dietary supplements containing isoflavones are also popular for treating various diseases, from postmenopausal disorders to cardiovascular diseases. Isoflavones are only synthesized in legumes, such as soy. In plants, isoflavones hold various important functions which govern plant survival and reproduction (222). The synthesis of the two major isoflavone structures, genistein and daidzein, is mediated by the membrane-bound protein cytochrome-P450 isoflavone synthase (IFS). The membrane-bound protein performs the oxidative attack and novel aryl ring migration of the flavanone substrates. Belonging to the type II eukaryotic cytochrome-P450, the catalysis of the microsomal IFS requires electrons which are transferred through the CPR partner proteins. The expression and catalysis of eukaryotic cytochrome-P450 enzymes cannot be supported by *E. coli* due to the lack of P450-redox partner proteins and incompatibility of membrane insertion signal particles. Therefore, heterologous expression of plant IFS has been done using yeast as a eukaryotic host (223). Recently, Ralston et al. (224) reported the partial reconstruction of isoflavonoid biosynthesis in *S. cerevisiae* by testing different isozymes of CHI and expressing them with IFS. Cultures fed with different chalcones (CHL) were able to produce flavanones and isoflavones when IFS was combined with CHI. Even though yeast as a recombinant platform demonstrated the feasibility of increasing isoflavonoid availability, the strategy still suffers from two disadvantages. The first concern was the low production yield, which could be attributed to the low turnover rate of the IFS enzymes in yeast. The second drawback was the necessity of chalcone supplementation into the yeast cultures as a precursor to isoflavones. In this case, since chalcones needed to be derived from chemical synthesis, one-pot isoflavone production could not be achieved. Indeed, this can be remedied through the engineering of the early pathway, as has been demonstrated for other functional flavonoid engineering endeavors in yeast (225) and by using a CPR/IFS fusion protein in *E. coli* (226).

3. Flavones

Flavones are a ubiquitous flavonoid class within the plant kingdom and contain important functions that affect plant survivability. This class of flavonoid is also of medicinal importance to promote human health for it was shown to induce apoptosis of HER2/neu breast cancer cells (227), to retard the growth of melanoma (228), and to inhibit the replication of human immunodeficiency virus type 1 (HIV-1) (229). However, despite its important health-promoting properties, flavones only occur in a small fraction of human diets, as the primary sources are in herbs such as parsley, thyme, and chamomile.

In order to increase flavone availability for nutraceutical or medicinal applications, recombinant strains of *E. coli* and baker's yeast have also been engineered to produce flavones. In most plants, flavones are synthesized from flavanones from the membrane-bound P450 flavone synthase II (FSII) (230). To avoid the hurdle of functional expression of plant P450s in *E. coli*, *S. cerevisiae* was used to construct flavonoid pathways. For example, the synthesis of the flavonoid naringenin from cinnamic acid was afforded by a recombinant *S. cerevisiae* strain expressing 4CL, CHS, CHI, and the P450 cinnamate 4-hydroxylase (C4H) that performs the hydroxylation of cinnamic acid into p-coumaric acid (231). This yeast strain was also adapted to produce flavones (225). In plants, the biosynthesis of flavones from flavanones is governed by two distinct enzymes. The soluble flavone synthase I (FSI) is found only among plants that belong to the Apiceae family (such as parsley), whereas the membrane-bound flavone synthase II (FSII) is ubiquitous in the plant kingdom. The existence of these two metabolic routes is intriguing. In order to investigate the efficiency of these two enzymes and the consequence of such diversion, FSI and FSII were functionally expressed in the flavanone producer yeast strain. To optimize FSII activity, the coexpression of CPR from yeast and *Catharanthus roseus* was evaluated. In vitro enzyme assays showed that FSII activity was most improved when *C. roseus* CPR was present. On the other hand, the coreactions with yeast CPR only improved flavone synthesis marginally. These results indicated that the interaction between cytochrome-P450s and their redox partners might be important in modulating reaction turnover rates. Metabolically engineered *S. cerevisiae* overexpressing FSI together with a flavanone biosynthetic pathway resulted in higher flavone production than the FSII-expressing recombinant strain (225). However, the higher synthesis of the flavone apigenin resulted in the reduced synthesis of the flavanone naringenin. Further studies indicated that apigenin could serve as a competitive inhibitor to the flavanone biosynthetic pathway. An additional flavonoid-modifying enzyme, 7-O-methyltransferase, yielded O-methylated flavones. Overall, the yeast system not only was useful as a P450 biocatalyst that can synthesize high-value flavonoids but also could be used as a model to investigate the metabolic reaction efficiency for the biosynthesis of plant flavonoids.

4. Flavanols

Flavonols are another class of flavonoids that diverge from the flavanone branch pathway. Flavanone 3β-hydroxylase (FHT) is the leading enzyme to convert flavanones into dihydroflavonols. Dihydroflavonols then serve as the substrates for another 2-oxoglutarate-dependent enzyme, flavonol synthase (FLS), to create flavonols. Similar to flavones, flavonols are of pharmacological importance as antimutagenic agents or as effective inhibitors of angiogenesis (232, 233). In order to obtain substantial production, *E. coli* was engineered for flavonol biosynthesis through the episomal expression of flavonol biosynthetic enzymes. In addition to expressing 4CL, CHS, and CHI, both *Malus domestica* FHT and *Arabidopsis thaliana* FLS were engineered in the flavanone-producing *E. coli* strain to result in the synthesis of naringenin, dihydrokaempferol, and kaempferol from *p*-coumaric acid (234). However, hydroxylated flavonoids (eriodictyol, dihydroquercetin, and quercetin) could not be detected in the culture medium when the flavonol-producing *E. coli* strain was supplemented with caffeic acid due to the low affinity of 4CL toward the dihydroxylated phenylpropanoid.

5. *Anthocyanins*

Anthocyanins are colorful phytochemicals that contribute to the red and blue colorations in plants which hold important functions to attract specific or nonspecific pollinators. In plants, anthocyanins are found attached to several other molecules, such as sugar groups and other flavonoids, to improve their stability and coloration (202, 235). For humans, anthocyanins have important pharmacological properties, as they serve, among others, as antioxidants (236) and cancer-preventive metabolites (237). Anthocyanins are available to humans through diets rich in colorful grains and fruits and vegetables and also red wine.

Several attempts have been made to produce and extract anthocyanins from plants with the purpose of utilizing them as colorants. The most abundant and historically oldest anthocyanin extract used as a natural colorant is derived from grape pomace (238). Others, either available commercially or used locally as natural extracts for coloring foodstuffs, include flower petals (239), grape rinds and red rice (240), red soybeans and red beans (241), *Vaccinium* species (242), purple corn (243), and different cherry extracts (244–246). However, in all cases, one of the biggest obstacles is the "browning effect." This refers to the formation of a brown color in plant anthocyanin extracts as a result of a two-step process. First, anthocyanins are oxidized by plant polyphenol oxidases present in the plant extract (247–249). Second, the oxidized anthocyanins undergo condensation and form brown pigments, which are usually undesired by the food industry. The red cabbage colorant is the only one that does not undergo browning and as a result is the most commonly used anthocyanin mixture.

As an alternative to plant extracts, bioreactor-based systems of plant cells for mass production of flavonoids, anthocyanins in particular, have been described for a few species (205, 206). However, economic feasibility has not been established yet, in part because of engineering challenges in mass cultivation of plant cultures. One challenge is that plant cells tend to form aggregates that influence anthocyanin culture productivity (250). Cells within aggregates are not adequately exposed to lighting required for flavonoid biosynthesis. For example, formation of phenylalanine ammonia lyase (PAL), a key enzyme in the biosynthetic pathway, is promoted primarily by UV wavelengths, particularly those of the UV-B region (251). Other enzymes in the pathway, particularly those of the anthocyanin biosynthetic branch, appear to be regulated in part by UV and in part by phytochrome-activating wavelengths (700–800 nm) (252). In that respect, irradiance becomes a limiting factor to productivity. This is because cells at the interior of an aggregate have limited or no exposure to light (253). Also, the average light dosage is reduced or insufficient within a dense cell culture since the cell wall composition selectively restricts certain wavelengths (254). On the other hand, flavonoid production in recombinant microorganisms is advantageous because the cloned pathway(s) are under microbial promoters and therefore the production is independent of light or other regulatory elements (such as the MYB transcription factors) required by plants. In addition, E. coli and S. cerevisiae cultures can achieve higher yields than plant cell cultures because they do not form aggregates and rapidly grow. As a result, these two organisms by and large remain the preferential production platforms in the biotechnology industry for several natural compounds and plant secondary metabolites (such as lycopene). As an added bonus, no plant peroxidases are present in bacteria and yeast and therefore the browning effect problem is significantly reduced. A simplified extraction procedure is another advantage of using microbial production platforms over plant crops or cultures. Since anthocyanins are not naturally produced in microbial hosts, a much less complicated matrix of products is generated through the heterologous expression of pathways that lead to specific product targets. This minimizes the downstream processing required for purification of the target molecules. These, as well as other issues inherent with plant cell suspension cultures (255), are some of the reasons why until now no such system has been established for the biosynthesis of anthocyanins, despite a more than 10-year effort in this field.

As a promising alternative to plant extracts and plant cell cultures, E. coli was engineered to produce anthocyanins (256). To achieve this goal, the flavanone pathway was bypassed by supplemental feeding of flavanones into E. coli JM109 carrying an assembly which consisted of M. domestica FHT, A. andraeanum DFR, anthocyanidin synthase (ANS) from apple, and UDP-glucose–flavonoid 3-O-glucosyltransferase (3GT) from Petunia. Unlike the production of other flavonoids, the synthesis of anthocyanin glucoside from the fermentation of E. coli supplemented with flavanones was very low.

Characterization of *A. thaliana* ANS activities in vitro indicated that the major products from the conversion of (*2R,3R,4S*)-leucoanthocyanidins were flavonols, with only minor anthocyanin aglycones. Recently, however, the role of ANS in the anthocyanin branch pathway was clarified through in vitro characterization of recombinant *Gerbera* ANS heterologously expressed in yeast. When (+)-catechin was added as the reaction substrate, the majority of the ANS reaction product was the 4,4-dimer of oxidized (+)-catechin and a trace amount of cyanidin and the flavonol quercetin. However, the addition of recombinant *Fragaria* 3GT derived from expression in *E. coli* resulted in conversion of (+)-catechin into cyanidin-3-*O*-glucoside as the major product and cyanidin and dimeric catechin as the minor products. Based on these results, (+)-catechin may be a better substrate for ANS, which could explain the minute production of anthocyanins from the engineered *E. coli*.

C. STILBENES

Resveratrol, a stilbene which is popularly associated with the health benefits of red wine, has recently attracted significant scientific attention. This simple flavonoid has been shown to extend the life span of the simple eukaryotic *S. cerevisiae* (257), *Drosophila melanogaster* (258), and *Caenorabitis elegans* (259) in a Sir2-dependent manner. Similarly, resveratrol exerted SIRT1-dependent activities which improved cellular functions and health. From these results, resveratrol could potentially be used to promote longevity in humans as well, since this compound also improved health and survival of mice on a high-calorie diet (260).

To increase the availability of this important stilbene, microorganisms have been engineered to serve as alternative production platforms for resveratrol production. Commonly utilized heterologous production platforms such as *S. cerevisiae* and *E. coli* do not naturally produce flavonoids. Therefore, the engineering of stilbene biosynthesis in these microorganisms requires grafting of plant biosynthetic pathways. The first endeavor of microbial flavonoid production was the metabolic engineering of resveratrol biosynthesis in *S. cerevisiae*. To enable resveratrol production from the *p*-coumaric acid precursor, the structural gene of 4CL from poplar and STS from grape were inserted into *S. cerevisiae* strain FY23. Expression of the plant 4CL and STS in FY23 was regulated under the control of the yeast alcohol dehydrogenase II promoter and the yeast enolase promoter, respectively. The resulting strain was capable of producing resveratrol for the first time (261). Metabolic engineering of resveratrol biosynthesis was also recently demonstrated in *E. coli* (262). In this case, *Arabidopsis thaliana* 4CL1 and peanut STS were introduced into *E. coli* strain BW27784. Expression of the plant enzymes under the *E. coli lac* promoter resulted in the synthesis of resveratrol from *p*-coumaric acid and piceatannol from caffeic acid. Additionally, 4CL from *Nicotiana tabacum* and STS from *Vinis vinifera* were both expressed in *E. coli* and *S. cerevisiae* with similar yields (263). Overall, these results have demonstrated the feasibility of synthesizing resveratrol compounds from genetically tractable microorganisms to increase the availability of this flavonoid.

D. P450-DERIVED AROMATICS

Cytochrome-P450 enzymes catalyze the regiospecific and stereospecific oxidation of nonactivated hydrocarbons at moderate temperatures. In plants, the P450 enzymes (type II) are bound to the membrane of the endoplasmic reticulum (ER) and require P450 reductase enzymes to shuttle electrons derived from the reduced form of nicotinamide adenine dinucleotide phosphate (NADPH) into the heme core of the P450s. Electrostatic fields generated by the P450s and the reductase proteins allow interactions between the two proteins to mediate electron transfer. An advantageous feature of utilizing yeast as a flavonoid production platform is the ability to naturally support the functional expression of eukaryotic P450 enzymes. As a eukaryote, yeasts contain ER and P450 redox partner proteins and are readily adaptable to accept P450s from higher organisms.

Escherichia coli is often the preferred heterologous production platform, not only because of its robust growth and ease of cultivation, but also because it is the most genetically tractable microbial host. Extensive characterization of *E. coli* has resulted in a wealth of available genetic manipulations that are useful for metabolic engineering purposes. However, functional expression of many eukaryotic proteins is hindered in *E. coli* due to, for example, codon incompatibility or posttranslational modifications essential for enzymatic activity. Such is the case for the functional expression in *E. coli* of the cytochrome-P450 flavonoid biosynthetic enzymes. The requirement of membrane attachment and an electron transfer system prohibits the heterologous functional expression of flavonoid P450 enzymes in *E. coli* because of the lack of ER and P450-redox partner proteins. Understanding the features of P450s is crucial to obtain their successful enzymatic expression in *E. coli*. Membrane-associated P450s are synthesized with an N-terminal leader sequence rich in hydrophobic residues that serve as recognition signal for translational insertion into the ER membrane (264). Adjacent to the hydrophobic-rich residue is a short stretch of positively charged amino acid residues which act as a stop signal of the translocation across the membrane (265). Moreover, a short proline-rich region that follows the signal-anchor sequence serves as the hinge between the N-terminal anchor and the cytoplasmic catalytic domain since proline is a well-known α-helix breaker (266). In the past, mutations in the proline cluster did not affect protein expression and membrane association was retained. However, carbon monoxide binding of the mutated proteins failed to display absorbance at 450 nm, the characteristic of P450-type proteins, which indicated the lack of proper heme incorporation into the expressed proteins due to improper folding (266).

Attempts to express membrane-bound P450 proteins in *E. coli* have been primarily geared for crystallography purposes, which is extremely complicated due to protein insolubility. So far, there are only three eukaryotic microsomal P450 enzymes that have been crystallized when using *E. coli* as a recombinant host (267–269). Removal of the transmembrane domain of various microsomal P450s has generally produced enzymes with low solubility that required detergents for solubilization. This indicates that there are regions within the protein body other than the membrane anchor module that are responsible for membrane association. Removal of the membrane-spanning N-terminus of the rabbit microsomal 2C5 caused release of the proteins from membranes following expression in *E. coli* in a high-ionic environment and without the need for detergent. However, aggregation was formed by the dissociated P450s, which was not conducive to crystallization. To minimize/abolish monofacial membrane interactions that caused aggregation, several amino acid substitutions were made within the protein body. The resulting modified microsomal P450 was soluble in a high-salt buffer and formed a monomer unit which was then successfully crystallized. In the case of the truncated 2C5, the catalytic activities could be reconstituted in vitro after the addition with rabbit NADPH P450 reductase, which indicated that the lack of substantial membrane attachment did not affect the catalysis of the proteins.

From the study on the expression of plant membrane–bound P450 limonene hydroxylase in *E. coli*, removal of the membrane-spanning region did not result in the detection of the P450 absorption peak at 450 nm (270). A similar result was also observed with the native, untruncated limonene hydroxylase. Furthermore, sodium dodecyl sulfate–polyacrylamide gel electrophoresis (SDS–PAGE) immunoblot analysis of the membrane fractions from *E. coli* expressing the truncated hydroxylase only detected a small amount of protein recognized by the rabbit polyclonal antibodies. This assay did not detect any protein in the membrane fractions of *E. coli* expressing the native hydroxylase, which indicated the absence of functional protein expression. The flavonoid P450 enzyme C4H has also been studied for expression in *E. coli*. In this case, it was found that immunoblots failed to detect the full-length protein as well as the construct with complete removal of the membrane anchor module after expression in various *E. coli* strains, indicating the lack of functional expression. It is interesting to observe the phenomenon of nonfunctional expression occurred in both cases of intact P450 and the N-terminally truncated proteins. The variations in sensitivity of N-terminal removal as a strategy for *E. coli* expression indicate a substantial structural divergence among microsomal P450 enzymes. Overall, studies of plant P450s revealed that the N-terminal hydrophobic region is required for the expression of the protein in *E. coli*. However, the native module seemed to lead to the synthesis of inactive, misfolded proteins that fail to incorporate heme into the active site. Moreover, high expression of the misfolded proteins exerted cellular damages to the *E. coli* host which resulted in cell lysis and reduced growth after protein induction (271).

Synthesis of misfolded proteins often occurs when RNA polymerase fails to recognize translation codons of the RNA transcripts. This phenomenon is especially prevalent when the RNA polymerase of a species organism is responsible for the translation of foreign proteins which originate from a different source organism. In the case of plant P450s, the N-terminal hydrophobic sequence contains signal specific for the recognition of ER membranes. Therefore, since ER does not exist in *E. coli*, it is not surprising that the N-terminal codons of the plant P450 are not compatible for expression in *E. coli*. Changes of several N-terminal codons of some plant P450s have been reported to remedy the expression problems. In the case of successful C4H expression in *E. coli*, the protein engineering method involved the combination of N-terminal truncation and codon replacements. Specifically, the first six codons encoding the hydrophobic membrane-spanning residues were deleted. The seventh codon, which encoded glutamic acid, was replaced by the *E. coli* ATG codon corresponding to methionine in order to allow translation initiation. The second codon of the truncated P450 encoding lysine residue (the eighth codon of the native P450) was changed into alanine because alanine is the preferred second codon for the *E. coli lacZ* gene (272). Moreover, the sixth codon of the truncated construct (the twelfth codon of the original protein), which encoded for glycine, was replaced with alanine to eliminate the amino acid that could serve as a helix breaker (271). Expression of the modified C4H in *E. coli* strain M15, DH5α, and JM109 did not result in the detection of any protein. However, expression of the modified construct in the minicell producing *E. coli* strain DS410 resulted in the synthesis of the modified P450 that could be isolated without solubilization steps or the addition of lipids. Even though the recombinant proteins isolated from DS410 membranes exhibited catalytic activity in vitro that was four- to sevenfold lower than in microsomes from induced plant cell, the reaction products could be detected through radiochromatography (271).

Successful expression of the P450 limonene hydroxylase was also achieved through modifications of the N-terminal membrane-anchor signal (270). In this case, similar to N-terminal truncation of C4H, the successful expression involved removal of six amino acid codons from the P450 hydroxylases. Following the codon removal, the truncated gene constructs were fused with a modified leader sequence of eight amino acids derived from CYP17A, the bovine P450 17α-hydroxylase (270). This construct seemed to provide good expression in *E. coli* strain JM109 but not in XL1Blue, and the catalysis of the protein products could be restored in vitro after the addition of P450 reductase. CYP17A was functionalized for *E. coli* expression through extensive modifications of the N-terminal codons (273). Specifically, the second codon which encoded for tryptophan was replaced with the *E. coli* codon for alanine since this codon is preferred for the expression of the *lacZ* gene (272) as mentioned previously. Nucleotide sequences for the fourth and fifth codons which both encode for leucine were replaced with another alanine-encoding codon (silent mutations) that is rich in adenosine and thymine nucleotides since this region of *E. coli* mRNAs were shown to be rich in adenosine and uridine (274). Two silent mutations were also made at the last nucleotides of the sixth and seventh codon by replacing with adenosine and thymine, respectively, to minimize messenger ribonucleic acid (mRNA) secondary-structure formation that can halt RNA polymerase processing (275).

Another route to synthesize hydroxylated flavonoids is through the oxidation process catalyzed by the flavonoid 3′-hydroxylase (F3′H) and flavonoid 3′,5′-hydroxylase (F3′5′H). Metabolic engineering hydroxylated flavonol biosynthesis in *E. coli* was achieved through protein engineering F3′5′H and grafting of the flavonol biosynthetic pathway (276). F3′5′H from *Catharanthus roseus* was chosen for the recombinant pathway because this enzyme exhibits broad substrate specificity and introduces both single and double hydroxyl groups into the flavonoid substrates. The synthesis of hydroxylated flavonoids required both the expression of the F3'5'H and the ability of the enzyme to catalyze the oxidation reaction in vivo. Specifically, four N-terminal codons were removed and the nucleotides of the fifth codon were replaced with ATG. Moreover, the second codon of the shortened F3'5'H which encodes for leucine was changed into alanine. To obtain enzymatic activities in vivo, the lack of P450-reductase enzymes in *E. coli* had to be compensated. For that reason, a shortened CPR gene derived from *Catharanthus* was introduced along with the modified F3'5'H as a functional translational fusion. CPR was fused with the modified F3'5'H through a short sequence that did not favor the formation of secondary structures which can interfere with the interactions between the two proteins. When the chimeric F3'5'H was coexpressed with the flavonol biosynthetic pathway, in the presence of *p*-coumaric acid, a small amount of the dihydroxylated flavonol, quercetin, could be recovered from the culture media, which indicated that the chimeric P450 was expressed and was catalytically active in *E. coli*. Supplemental feeding of the flavanone naringenin to the recombinant *E. coli* strain resulted in the synthesis of dihydrokaempferol and kaempferol as well as the di- and trihydroxylated flavonoids dihydroquercetin, quercetin, and myricetin. Even though protein engineering of the F3'5'H resulted in the functional expression of the enzyme in *E. coli*, the low amount of the hydroxylated products indicated that the engineered enzyme did not attain optimum activities (276).

Moreover, it was also reported that some cell lysis occurred after induction of protein expression, which presents evidence that the expression of the proteins was toxic to the cells. In the metabolic engineering strategy, the toxicity of the P450 proteins was minimized through the placement of the gene into a low-copy-number expression plasmid to reduce protein production. Another important parameter for successful P450 expression in *E. coli* is the use of low- to medium-strength tightly regulated promoter. The expression of P450s in *E. coli* under the regulation of strong promoters appeared to favor the formation of inclusion bodies. The use of tightly regulated promoters is also especially important for toxic protein expression to prevent early protein synthesis, which occurs when using "leaky" promoters. In general, this study demonstrated that the synthesis of hydroxylated products need not rely on more complicated organisms, such as yeast or insect cells. *Escherichia coli*, a commonly used and robust biocatalyst, could be engineered for hydroxylated product synthesis through the functionalization of microsomal enzymes.

VI. Biofuels

The utility of engineered microorganisms as biocatalysts has been extended from fine-chemical to fuel synthesis. Although recently a handful of biofuel molecules can be synthesized from cell culture and fermentation technology, ethanol and butanol are the two alcohols whose microbial synthesis has met commercial productivity. Ethanol synthesis traditionally relies on yeast fermentation. However, through extensive genetic engineering, *E. coli* can now be used as a competitive biocatalyst. The synthesis of butanol also utilized bacterial fermentation. In this case, the solvent-producing gram-positive bacteria of the genus *Clostridium* were effectively used. Butanol fermentation technology was abandoned due to the development of efficient synthetic methods for butanol production. However, the recent movement of green technology has resurrected butanol synthesis through microbial processes. In general, there are still many challenges facing the use of microorganisms for biofuel synthesis. In this section, the various efforts to improve biofuel production through biocatalytic technology are covered.

A. ETHANOL

According to the U.S. International Trade Commission, 4.85 billion gallons of ethanol were produced in the United States in 2006, an increase of over 3 billion from only 5 years prior and 4 billion over the last 10 (277, 278). To this end, several microorganisms have been utilized for the conversion of biomass to ethanol. Initially utilized in food processing, *S. cerevisiae* is now actively used in fermentation for ethanol fuel synthesis. The bacterium *Zymomonas mobilis* is also naturally ethanolgenic and in some cases is more robust than yeast for bioethanol production. A major disadvantage of *S. cerevisiae* and *Z. mobilis* is the inability to ferment pentose sugars, the abundant fraction of hemicellulose biomass. On the other hand, *E. coli* can readily utilize both hexose and pentose sugars (279). Ethanol synthesis from *E. coli* is not substantial, and for this reason a variety of *E. coli* strains have been engineered to afford high-level ethanol synthesis (280). The successful engineering of high-level ethanol biosynthesis from *E. coli* was accomplished by implantation of the *Z. mobilis* homoethanol pathway (281–283). Ethanol biosynthesis (Figure 9A) in *Z. mobilis* starts from the degradation of pyruvate into acetaldehyde by the action of pyruvate decarboxylase (*pdc*). Subsequently, acetaldehyde is converted into ethanol by alcohol dehydrogenase (*adh*). To engineer ethanolgenic *E. coli*, the *Z. mobilis*

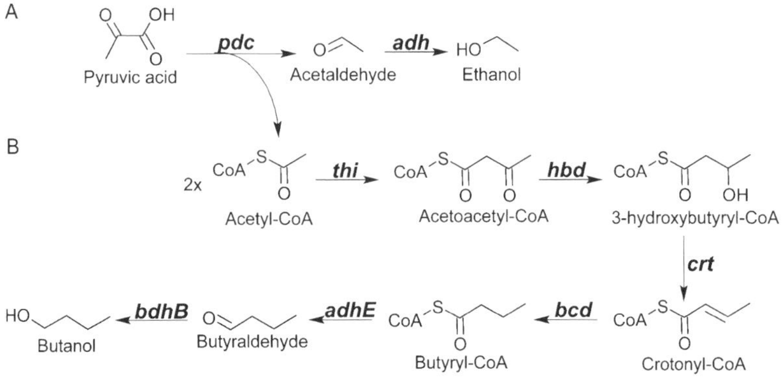

Figure 9.   (A) The *E. coli* ethanol pathway for the synthesis of ethanol. (B) *Clostridia* butanol pathway for the synthesis of butanol. The genes encoding for enzymes, indicated in bold, are: *pcd*: pyruvate decarboxylase; *adh*: alcohol dehydrogenase; *thi*: thiolase; *hbd*: 3-hydroxybutyryl-CoA dehydrogenase; *crt*: crotonase; *bcd*: butyryl-CoA dehydrogenase; *adhE*: butyraldehyde dehydrogenase; *bdhB*: butanol dehydrogenase.

homoethanol pathway was inserted into *E. coli* while simultaneously, to reduce carbon consumption by competing pathways, the gene encoding fumarate reductase (*frd*) in the ethanolgenic *E. coli* was mutated to inactivate succinate production. The rate of ethanol synthesis from the engineered *E. coli* (named KO11) was as high as yeast (284). The robustness of KO11 for ethanol production from biomass has also been demonstrated. The substrates tested included rice hull, sugar cane begasse, corn hulls and fibers, beet pulp, corn hulls, pine soft wood, orange peel, sweet whey, willow (hardwood), and brewery wastewater (285). *Klebsiella oxytoca* is an ethanolgenic bacterium that can metabolize a variety of biomass monomeric sugars. In comparison with *K. oxytoca* P2, *E. coli* KO11 could achieve higher ethanol productivities using most of the biomass substrates. For example, fermentation of sugar cane bagasse using KO11 reached 90% theoretical yield (37 g/L ethanol in 60 h), while fermentation using *K. oxytoca* P2 only reached 70% (39 g/L ethanol in 168 h) (285).

Despite the successful development of microbial biocatalysts, ethanol synthesis through *E. coli* fermentation suffers from several drawbacks. One major drawback is caused by the toxicity of fermentative products, which hampers the sustainability of the microbial biocatalysts to reach high-level productivity. On this note, even though the production rate of *E. coli* KO11 was as high as yeast, it exhibited lower ethanol tolerance than yeast (lack of growth in the presence of 35 g/L ethanol) (286). Strategies to improve ethanol tolerance in microorganisms typically involved media formulation, adaptation, and mutant selection (287). For example, in order to develop increased ethanol tolerance, KO11 was slowly adapted to higher ethanol exposure by serial subculturing. Over the course of three months, KO11 mutants were enriched in liquid media containing increased ethanol concentration and isolated on solid media. At the end of the metabolic evolution, a mutant (LY01) which could grow in the presence of 50 g/L ethanol could be found. LY01 had greater than 80% survival from brief exposure to 100 g/L ethanol, compared to only 10% survival for KO11 (286). The transcription profiles of KO11 and LY01 grown in rich medium with glucose or xylose with 0, 10, and 20 g/L ethanol were examined. Approximately 200 genes were regulated differently in LY01 when compared to KO11. These differentially expressed genes were involved in amino acid biosynthesis, cell processes, cell structure, central metabolism, energy metabolism, and stress-related responses. Three major differences between LY01 and KO11 were noted: increased expression of genes involved in betaine biosynthesis and glycine degradation, improvement in protective osmolyte uptake, and lack of fumarate and nitrate reductase regulator (FNR) function (288). Glycine metabolism and expression of FNR-regulated genes are responsible for regulating pyruvate availability. Betaine biosynthesis is also controlled by FNR through the ArcA regulon. All together, it was concluded that increased ethanol tolerance in LY01 appeared to result from the combination of differential pyruvate distribution and increased osmotic protection (288).

This method for the development of ethanol-tolerant *E. coli* strains, while successful, is time consuming. Thus, a novel technology was recently developed to engineer increased ethanol tolerance in *E. coli* rapidly by tuning global gene expression. To search for tolerant phenotypes, through tuning multiple genes, directed evolution was applied to a global regulon, the *rpoD* gene, which encodes for sigma 70 (289). The sigma factor was chosen as a target since it was hypothesized that beneficial mutations can alter the promoter preferences of RNA polymerase and affect the transcription levels of multiple genes. In this work, the *E. coli rpoD* gene and the upstream, intergenic promoter region was subjected to polymerase chain reaction (PCR) random mutagenesis and cloned into a low-copy-number expression vector. Phenotypic screening and selection were performed on 106 transformants. Specifically, transformants that grew on agar plates were initially grown in liquid. Samples from the liquid library were subcultured twice in challenging environments (supplementation of 50 g/L ethanol) to select for surviving mutants. The liquid library was then plated in order to facilitate the selection of 20 colonies that were subsequently assayed for growth in varying ethanol concentrations. To confirm that the improved phenotype was derived from mutation of the plasmid-borne sigma factor and not the chromosomal copy, plasmid DNA was isolated from the mutant strains and used to transform fresh, wild-type *E. coli*. For verification, a growth phenotype of the newly transformed strain was assayed in challenging environments. After confirming that indeed the improved phenotype was conferred by mutations in the plasmid-borne gene, the best mutant sigma factor was subjected to two additional rounds of mutation and selection. In both subsequent rounds, ethanol concentration was increased to 60 and 70 g/L in the selection stage. Various point mutations were identified across the sigma factor gene. After three iterative directed evolutions, the doubling time of the mutant strain was approximately 25-fold faster than the wild-type strain in ethanol-containing medium. Through transcriptional analysis of ethanol-tolerant mutants, it could be observed that the mutation copy of sigma factor altered various global gene expression patterns when compared to the wild-type copy.

A similar approach was performed to elicit ethanol-tolerant phenotypes in yeast. The transcription machinery in a eukaryotic cell is more complex than in a prokaryotic ell such as *E. coli*. For example, in *E. coli*, there is only one RNA polymerase, whereas in yeast, transcription is performed by three separate enzymes. To elicit a differential transcriptional phenotype in yeast, the TATA-binding protein (SPT15), one of the components of the general RNA Pol II transcription factor D was subjected to error-prone PCR based on the finding that mutations in a TATA-binding protein changed the preference of the there polymerases and played important roles in promoter specificity. In parallel, the effect of mutations on one associated factor, the TAF25 protein, was also investigated. Initially library screening of mutant SPT15 showed modest growth in the presence of 5% ethanol and 100 g/L glucose. To isolate dominant mutants, the library was subjected to increased ethanol concentration of 6% and 120 g/L glucose. Mutants that exhibited colony-forming ability on agar plate were selected for sequence analysis. Sequence analysis of the mutant factors which conferred to the best growth phenotype under ethanol and glucose stress revealed the presence of three mutations. For spt15, these mutations were localized in the second repeat element consisting of β-sheets. Mutations created in TAF25 were widespread throughout the gene. While mutant SPT15 provided 13-fold growth improvement in growth yield in some challenging conditions, it was concluded that mutant TAF25 was unable to grow in the presence of 6% ethanol. To analyze the effect of individual mutations in the identified triple mutant, all possible single and double mutations were re-created in SPT15. Assayed for ethanol and glucose tolerance phenotype, it was shown that none of the single or double mutations conferred tolerance at the level of the triple mutant (289).

It was suggested that mutations in genes of different transcription machinery resulted in different phenotypic responses. Through transcriptional profiling, it was demonstrated that the widespread alteration of many genes in yeast caused by mutant SPT15 was similar to that observed in *E. coli* with altered sigma factor. In *E. coli*, alteration of sigma factor resulted in a similar distribution between upregulated and downregulated genes. However in yeast, the majority of the genes with altered expression are upregulated. To determine if the increased ethanol and glucose tolerance was a result of upregulation of individual genes, individual gene knockouts were performed. For this purpose, 12 of the most highly expressed genes in the mutant under unstressed condition (0% ethanol and 20 g/L glucose) were selected, along with two additional genes. This experiment showed that deletion of the majority of these gene targets resulted in loss of the mutant SPT15 to generate increased ethanol and glucose tolerance. Without the overexpression of mutant SPT15, all knockout mutants exhibited normal tolerance level for ethanol and glucose. To further investigate if individual genes that were upregulated in the SPT15 mutant could elicit improved phenotype, three genes that were greatly upregulated were overexpressed (289). The results showed that the overexpression of individual genes could not elicit ethanol and glucose tolerance. All together, the data suggested that in SPT15 mutant many genes that were differentially regulated worked in concert in order display a novel phenotype and that individual genes were insufficient to exhibit increased ethanol and glucose tolerance.

B. BUTANOL

Due to the physical properties of butanol, the four-carbon alcohol is a better replacement for gasoline than ethanol. Unlike ethanol, the hydrophobicity of butanol prevents water contamination. Additionally, the vapor pressure of butanol (4 mm Hg at 20°C) is much lower than ethanol (45 mm Hg at 20°C). As such, these two properties allow the utilization of existing infrastructure for butanol storage and transportation. Butanol also has a similar energy density to gasoline (butanol 27 MJ/L; gasoline 32 MJ/L), and it can be used as a direct substitute for gasoline.

Various clostridia have been utilized in butanol fermentation, although the gram-positive anaerobe produces butanol yield contaminated with by-products butyric acid, acetone, and ethanol, complicating production (290). Clostridia also grow slowly and their spore-forming life cycle complicates industrial application. From a biotechnological perspective, the lack of genetic characterization hinders metabolic or genetic engineering endeavors for butanol synthesis optimization and by-product reduction. Because of these reasons, recently, *E. coli* was engineered to afford butanol biosynthesis from sugar (291). The engineering strategy involved the heterologous expression of *Clostridium acetobutylum* butanol biosynthetic enzymes (Figure 9B). In clostridia, two molecules of acetyl-CoA are condensed into acetoacetyl-CoA by the enzyme thiolase (*thi*). Subsequently, 3-hydroxybutyryl-CoA is formed by the action of hydroxybutyryl-CoA dehydrogenase (*hbd*). Next, the enzyme crotonase (*Ca-crt*) catalyzes the formation of crotonyl-CoA from 3-hydroxybutyryl-CoA. Crotonoyl-CoA is then converted into butyryl-CoA by butyryl-CoA dehydrogenase (*bcd*) and electron transfer proteins (*etfAB*). Finally, butyraldehyde and butanol are formed by the action of butyraldehyde dehydrogenase (*Ca-adhE*) and butanol dehydrogenase (*bdhB*). In *C. acetobutylum*, genes *hbd*, *Ca-crt*, *bcd*, and *etfAB* are organized into a single operon, while the genes *thi*, *Ca-adhE*, and *bdhB* are located separately in the genome. The metabolic engineering of butanol biosynthesis in *E. coli* was performed by cloning the butanol biosynthetic operon in an *E. coli* expression plasmid under the control of the IPTG-inducible PLlacO1 promoter. Genes *adhE*, *bdhB*, and *thi* were placed together under another coreplicable plasmid under the control of the same promoter. Initial anaerobic fermentation of the recombinant strain resulted in around 14 mg/L butanol synthesis. By slightly increasing oxygen exposure, butanol production could be slightly improved, which suggested that the NADH cofactor required for butanol biosynthesis was insufficiently generated under anaerobic condition. When cultured in complete aerobic condition, butanol synthesis decreased since in *E. coli* both acetyl-CoA and NADH are metabolized in the tricarboxylic acid (TCA) cycle and respiration.

Evidently, expression of the butanol pathway only resulted in low butanol synthesis. In order to improve productivity, some endogenous *E. coli* pathways were deleted to alleviate metabolic drain such as acetyl-CoA and NADH. Since deletion targets of endogenous metabolism are often less predictable, several deletion schemes were investigated. When the native *E. coli* genes encoding for lactate dehydrogenase (*ldhA*), alcohol dehydrogenase (*Ec-adhE*), and fumarate reductase (*frdBC*) were deleted, butanol production improved two fold, in parallel to the reduction of undesired fermentative products such as lactate, ethanol, and succinate. Using this genetic background, acetate production was also increased. In order to reduce acetate by-product, the gene *pta* encoding for phosphotransacetylase was also removed. The four-deletion resulted in lower acetate accumulation; however, the synthesis of butanol also decreased. Another deletion, that of the gene *fnr*, which encoded for an anaerobic regulator, was intended to relieve the repression of pyruvate dehydrogenase. However, low butanol production was obtained when *fnr* was deleted in conjunction with *ldhA*, *Ec-adhE*, and *frdBC*. In contrast, butanol synthesis improved almost three fold relative to that of the wild-type level, when both *fnr* and *pta* were deleted in *ldhA*, *Ec-adhE*, and *frdBC* deletion background. In this deletion scheme, ethanol and pyruvate levels were also increased. All together, these deletion studies showed that the deletion of *fnr* alone did not fully increase the capacity of pyruvate dehydrogenase. Moreover, the deterministic approaches to control endogenous fluxes for butanol production improvement appeared to result in unpredictable outcome. Combining all of the optimization strategies, the maximum butanol production of the engineered *E. coli* reached approximately 500 mg/L in rich medium fermentation. When compared to the current state of *Clostridium* fermentation, the *E. coli* fermentation is approximately 30-fold lower in butanol yield. However, the wealth of genetic tools developed for manipulating the genetics and metabolism in *E. coli* could lead to the development of competitive or better heterologous biocatalyst for butanol production.

VII.   Summary

Ever since the era of recombinant DNA technology for natural product biosynthesis emerged (292), microorganisms are increasingly becoming common production platforms for many fine chemicals, including natural products and biofuels, that are currently being produced either through chemical methods or using plant and organ cell cultures. The rapid elucidation of biosynthetic pathways made possible through advanced genomic tools has made natural products again the molecules of choice for drug development. Indeed, half of the drugs currently in clinical use are natural products and it is expected that the market size of biotechnology-derived small molecules will exceed billion U.S.$100 in 2010 and billion U.S.$400 in 2030 (3, 293). There are still many challenges facing the use of microorganisms for high-value chemical synthesis. For example, further developments of recent advances are necessary to make a fermentation-based biobutanol industry that can compete effectively with petrochemically derived butanol. As such, we believe that biocatalyst factories such as *E. coli* and *S. cerevisiae* will not only continue to be highly attractive alternatives to traditional chemical manufacturing but the application of powerful systems biology approaches will facilitate their expanded role in industrial applications (294–296).

References

1. Maimone, T. J., and Baran, P. S. (2007) Modern synthetic efforts toward biologically active terpenes, *Nat. Chem. Biol. 3*, 396–407.

2. Roberts, S. C. (2007) Production and engineering of terpenoids in plant cell culture, *Nat. Chem. Biol. 3*, 387–395.

3. Maury, J., Asadollahi, M. A., Moller, K., Clark, A. M., and Nielsen, J. (2005) Microbial isoprenoid production: An example of green chemistry through metabolic engineering, *Ad. Biochem. Eng./Biotechnol. 100*, 19–51.

4. Winkler, J. D., Rouse, M. B., Greaney, M. F., Harrison, S. J., and Jeon, Y. T. (2002) The first total synthesis of (+/−)-ingenol, *J. Am. Chem. Soc. 124*, 9726–9728.

5. Wender, P. A., Jesudason, C. D., Nakahira, H., Tamura, N., Tebbe, A. L., and Ueno, Y. (1997) The first synthesis of a daphnane diterpene: The enantiocontrolled total synthesis of (+)-resiniferatoxin, *J. Am. Chem. Soc. 119*, 12976–12977.

6. Mandal, M., Yun, H., Dudley, G. B., Lin, S., Tan, D. S., and Danishefsky, S. J. (2005) Total synthesis of guanacastepene a: A route to enantiomeric control, *J. Org. Chem. 70*, 10619–10637.

7. Watts, K. T., Mijts, B. N., and Schmidt-Dannert, C. (2005) Current and emerging approaches for natural product biosynthesis in microbial cells, *Adv. Synth. Catal. 347*, 927–940.

8. Kim, H. J., and Lee, I. S. (2006) Microbial metabolism of the prenylated chalcone xanthohumol, *J. Nat. Prod. 69*, 1522–1524.

9. Picaud, S., Mercke, P., He, X., Sterner, O., Brodelius, M., Cane, D. E., and Brodelius, P. E. (2006) Amorpha-4,11-diene synthase: Mechanism and stereochemistry of the enzymatic cyclization of farnesyl diphosphate, *Arch. Biochem. Biophys. 448*, 150–155.

10. Goldstein, J. L., and Brown, M. S. (1990) Regulation of the mevalonate pathway, *Nature 343*, 425–430.

11. Hampton, R., Dimster-Denk, D., and Rine, J. (1996) The biology of HMG-CoA reductase: The pros of contra-regulation, *Trends Biochem. Sci. 21*, 140–145.

12. Hampton, R. Y. (1998) Genetic analysis of hydroxymethylglutaryl-coenzyme A reductase regulated degradation, *Curr. Opin. Lipidol. 9*, 93–97.

13. Thorsness, M., Schafer, W., D'Ari, L., and Rine, J. (1989) Positive and negative transcriptional control by heme of genes encoding 3-hydroxy-3-methylglutaryl coenzyme A reductase in Saccharomyces cerevisiae, *Mol. Cell. Biol. 9*, 5702–5712.

14. Turi, T. G., and Loper, J. C. (1992) Multiple regulatory elements control expression of the gene encoding the Saccharomyces cerevisiae cytochrome P450, lanosterol 14 alpha-demethylase (ERG11), *J. Biol. Chem. 267*, 2046–2056.

15. Bach, T. J., Boronat, A., Campos, N., Ferrer, A., and Vollack, K. U. (1999) Mevalonate biosynthesis in plants, *Crit. Rev. Biochem. Mol. Biol. 34*, 107–122.

16. Daum, G., Lees, N. D., Bard, M., and Dickson, R. (1998) Biochemistry, cell biology and molecular biology of lipids of Saccharomyces cerevisiae, *Yeast 14*, 1471–1510.

17. Dorsey, J. K., and Porter, J. W. (1968) The inhibition of mevalonic kinase by geranyl and farnesyl pyrophosphates, *J. Biol. Chem. 243*, 4667–4670.

18. Rohmer, M., Knani, M., Simonin, P., Sutter, B., and Sahm, H. (1993) Isoprenoid biosynthesis in bacteria: a novel pathway for the early steps leading to isopentenyl diphosphate, *Biochem. J. 295*(Pt. 2), 517–524.

19. Rodriguez-Concepcion, M. (2004) The MEP pathway: A new target for the development of herbicides, antibiotics and antimalarial drugs, *Curr. Pharm. Des. 10*, 2391–2400.

20. Takahashi, S., Kuzuyama, T., Watanabe, H., and Seto, H. (1998) A 1-deoxy-D-xylulose 5-phosphate reductoisomerase catalyzing the formation of 2-C-methyl-D-erythritol 4-phosphate in an alternative nonmevalonate pathway for terpenoid biosynthesis, *Proc. Natl. Acad. Sci. USA 95*, 9879–9884.

21. Kuzuyama, T., Takagi, M., Takahashi, S., and Seto, H. (2000) Cloning and characterization of 1-deoxy-D-xylulose 5-phosphate synthase from Streptomyces sp. strain CL190, which uses both the mevalonate and nonmevalonate pathways for isopentenyl diphosphate biosynthesis, *J. Bacteriol. 182*, 891–897.

22. Kuzuyama, T., Takahashi, S., Takagi, M., and Seto, H. (2000) Characterization of 1-deoxy-D-xylulose 5-phosphate reductoisomerase, an enzyme involved in isopentenyl diphosphate biosynthesis, and identification of its catalytic amino acid residues, *J. Biol. Chem. 275*, 19928–19932.

23. Takagi, M., Kuzuyama, T., Takahashi, S., and Seto, H. (2000) A gene cluster for the mevalonate pathway from Streptomyces sp. strain CL190, *J. Bacteriol. 182*, 4153–4157.

24. Herz, S., Wungsintaweekul, J., Schuhr, C. A., Hecht, S., Luttgen, H., Sagner, S., Fellermeier, M., Eisenreich, W., Zenk, M. H., Bacher, A., and Rohdich, F. (2000) Biosynthesis of terpenoids: YgbB protein converts 4-diphosphocytidyl-2C-methyl-D-erythritol 2-phosphate to 2C-methyl-D-erythritol 2,4-cyclodiphosphate, *Proc. Natl. Acad. Sci. USA 97*, 2486–2490.

25. Luttgen, H., Rohdich, F., Herz, S., Wungsintaweekul, J., Hecht, S., Schuhr, C. A., Fellermeier, M., Sagner, S., Zenk, M. H., Bacher, A., and Eisenreich, W. (2000) Biosynthesis of terpenoids: YchB protein of Escherichia coli phosphorylates the 2-hydroxy group of 4-diphosphocytidyl-2C-methyl-D-erythritol, *Proc. Natl. Acad. Sci. USA 97*, 1062–1067.

26. Rohdich, F., Wungsintaweekul, J., Eisenreich, W., Richter, G., Schuhr, C. A., Hecht, S., Zenk, M. H., and Bacher, A. (2000) Biosynthesis of terpenoids: 4-Diphosphocytidyl-2C-methyl-D-erythritol synthase of Arabidopsis thaliana, *Proc. Natl. Acad. Sci. USA 97*, 6451–6456.

27. Rohdich, F., Wungsintaweekul, J., Fellermeier, M., Sagner, S., Herz, S., Kis, K., Eisenreich, W., Bacher, A., and Zenk, M. H. (1999) Cytidine 5′-triphosphate-dependent

biosynthesis of isoprenoids: YgbP protein of Escherichia coli catalyzes the formation of 4-diphosphocytidyl-2-C-methylerythritol, *Proc. Natl. Acad. Sci. USA 96*, 11758–11763.

28. Rohdich, F., Wungsintaweekul, J., Luttgen, H., Fischer, M., Eisenreich, W., Schuhr, C. A., Fellermeier, M., Schramek, N., Zenk, M. H., and Bacher, A. (2000) Biosynthesis of terpenoids: 4-Diphosphocytidyl-2-C-methyl-D-erythritol kinase from tomato, *Proc. Natl. Acad. Sci. USA 97*, 8251–8256.

29. Campos, N., Rodriguez-Concepcion, M., Sauret-Gueto, S., Gallego, F., Lois, L. M., and Boronat, A. (2001) Escherichia coli engineered to synthesize isopentenyl diphosphate and dimethylallyl diphosphate from mevalonate: A novel system for the genetic analysis of the 2-C-methyl-d-erythritol 4-phosphate pathway for isoprenoid biosynthesis, *Biochem. J. 353*, 59–67.

30. Fellermeier, M., Raschke, M., Sagner, S., Wungsintaweekul, J., Schuhr, C. A., Hecht, S., Kis, K., Radykewicz, T., Adam, P., Rohdich, F., Eisenreich, W., Bacher, A., Arigoni, D., and Zenk, M. H. (2001) Studies on the nonmevalonate pathway of terpene biosynthesis. The role of 2C-methyl-D-erythritol 2,4-cyclodiphosphate in plants, *Eur. J. Biochem. 268*, 6302–6310.

31. Hecht, S., Eisenreich, W., Adam, P., Amslinger, S., Kis, K., Bacher, A., Arigoni, D., and Rohdich, F. (2001) Studies on the nonmevalonate pathway to terpenes: The role of the GcpE (IspG) protein, *Proc. Natl. Acad. Sci. USA 98*, 14837–14842.

32. Kemp, L. E., Bond, C. S., and Hunter, W. N. (2002) Structure of 2C-methyl-D-erythritol 2,4-cyclodiphosphate synthase: An essential enzyme for isoprenoid biosynthesis and target for antimicrobial drug development, *Proc. Natl. Acad. Sci. USA 99*, 6591–6596.

33. Lehmann, C., Lim, K., Toedt, J., Krajewski, W., Howard, A., Eisenstein, E., and Herzberg, O. (2002) Structure of 2C-methyl-D-erythrol-2,4-cyclodiphosphate synthase from Haemophilus influenzae: Activation by conformational transition, *Proteins 49*, 135–138.

34. Richard, S. B., Ferrer, J. L., Bowman, M. E., Lillo, A. M., Tetzlaff, C. N., Cane, D. E., and Noel, J. P. (2002) Structure and mechanism of 2-C-methyl-D-erythritol 2,4-cyclodiphosphate synthase. An enzyme in the mevalonate-independent isoprenoid biosynthetic pathway, *J. Biol. Chem. 277*, 8667–8672.

35. Rohdich, F., Eisenreich, W., Wungsintaweekul, J., Hecht, S., Schuhr, C. A., and Bacher, A. (2001) Biosynthesis of terpenoids. 2C-Methyl-D-erythritol 2,4-cyclodiphosphate synthase (IspF) from Plasmodium falciparum, *Eur. J. Biochem. 268*, 3190–3197.

36. Steinbacher, S., Kaiser, J., Wungsintaweekul, J., Hecht, S., Eisenreich, W., Gerhardt, S., Bacher, A., and Rohdich, F. (2002) Structure of 2C-methyl-D-erythritol-2,4-cyclodiphosphate synthase involved in mevalonate-independent biosynthesis of isoprenoids, *J. Mol. Biol. 316*, 79–88.

37. Seemann, M., Tse Sum Bui, B., Wolff, M., Miginiac-Maslow, M., and Rohmer, M. (2006) Isoprenoid biosynthesis in plant chloroplasts via the MEP pathway: Direct thylakoid/ferredoxin-dependent photoreduction of GcpE/IspG, *FEBS Lett. 580*, 1547–1552.

38. Zepeck, F., Grawert, T., Kaiser, J., Schramek, N., Eisenreich, W., Bacher, A., and Rohdich, F. (2005) Biosynthesis of isoprenoids. Purification and properties of IspG protein from Escherichia coli, *J. Org. Chem. 70*, 9168–9174.

39. Altincicek, B., Kollas, A., Eberl, M., Wiesner, J., Sanderbrand, S., Hintz, M., Beck, E., and Jomaa, H. (2001) LytB, a novel gene of the 2-C-methyl-D-erythritol 4-phosphate pathway of isoprenoid biosynthesis in Escherichia coli, *FEBS Lett 499*, 37–40.

40. Cunningham, F. X., Jr., Lafond, T. P., and Gantt, E. (2000) Evidence of a role for LytB in the nonmevalonate pathway of isoprenoid biosynthesis, *J. Bacteriol. 182*, 5841–5848.

41. Frense, D. (2007) Taxanes: Perspectives for biotechnological production, *Appl. Microbiol. Biotechnol. 73*, 1233–1240.

42. Lin, L., and Wu, J. (2002) Enhancement of shikonin production in single- and two-phase suspension cultures of Lithospermum erythrorhizon cells using low-energy ultrasound, *Biotechnol. Bioeng. 78*, 81–88.

43. Zhong, J. J. (2001) Biochemical engineering of the production of plant-specific secondary metabolites by cell suspension cultures, *Adv. Biochem. Eng. Biotechnol. 72*, 1–26.

44. Balandrin, M. F., Klocke, J. A., Wurtele, E. S., and Bollinger, W. H. (1985) Natural plant chemicals: Sources of industrial and medicinal materials, *Science 228*, 1154–1160.

45. Srivastava, S., and Srivastava, A. K. (2007) Hairy root culture for mass-production of high-value secondary metabolites, *Crit. Rev. Biotechnol. 27*, 29–43.

46. Gibson, D. M., Ketchum, R. E. B., Vance, N. C., and Christen, A. A. (1993) Initiation and growth of cell-lines of Taxus-brevifolia (Pacific yew), *Plant Cell Rep. 12*, 479–482.

47. Linden, J. C., Mirjalili, N., Haigh, J. R., and Sun, X. (1995) An examination of dissolved-gas effects on growth and secondary metabolism of Taxus and Artemisia cell-cultures, *Abstr. Pap. Am. Chem. Soc. 209*, 82.

48. Mirjalili, N., and Linden, J. C. (1995) Gas-phase composition effects on suspension-cultures of Taxus-Cuspidata, *Biotechnol. Bioeng. 48*, 123–132.

49. Mirjalili, N., and Linden, J. C. (1996) Methyl jasmonate induced production of taxol in suspension cultures of Taxus cuspidata: Ethylene interaction and induction models, *Biotechnol. Prog. 12*, 110–118.

50. Ketchum, R. E. B., Gibson, D. M., Croteau, R. B., and Shuler, M. L. (1999) The kinetics of taxoid accumulation in cell suspension cultures of Taxus following elicitation with methyl jasmonate, *Biotechnol. Bioeng. 62*, 97–105.

51. Hezari, M., Ketchum, R. E. B., Gibson, D. M., and Croteau, R. (1997) Taxol production and taxadiene synthase activity in Taxus canadensis cell suspension cultures, *Arch. Bioch. Biophys. 337*, 185–190.

52. Ketchum, R. E. B., Tandon, M., Gibson, D. M., Begley, T., and Shuler, M. L. (1999) Isolation of labeled 9-dihydrobaccatin III and related taxoids from cell cultures of Taxus canadensis elicited with methyl jasmonate, *J. Nat. Prod. 62*, 1395–1398.

53. Gueritte, F. (2001) General and recent aspects of the chemistry and structure-activity relationships of taxoids, *Curr. Pharm. Des. 7*, 1229–1249.

54. Kingston, D. G. (1991) The chemistry of taxol, *Pharmacol. Ther. 52*, 1–34.

55. Shuler, M. L. (1994) Bioreactor engineering as an enabling technology to tap biodiversity. The case of taxol, *Ann. NY Acad. Sci. 745*, 455–461.

56. Qian, Z. G., Zhao, Z. J., Xu, Y., Qian, X., and Zhong, J. J. (2005) Highly efficient strategy for enhancing taxoid production by repeated elicitation with a newly synthesized

jasmonate in fed-batch cultivation of Taxus chinensis cells, *Biotechnol. Bioeng. 90*, 516–521.

57. Qian, Z. G., Zhao, Z. J., Xu, Y., Qian, X., and Zhong, J. J. (2005) A novel synthetic fluoro-containing jasmonate derivative acts as a chemical inducing signal for plant secondary metabolism, *Appl. Microbiol. Biotechnol. 68*, 98–103.

58. Qian, Z. G., Zhao, Z. J., Xu, Y., Qian, X., and Zhong, J. J. (2006) Novel synthetic 2,6-dichloroisonicotinate derivatives as effective elicitors for inducing the biosynthesis of plant secondary metabolites, *Appl. Microbiol. Biotechnol. 71*, 164–167.

59. Pyo, S. H., Choi, H. J., and Han, B. H. (2006) Large-scale purification of 13-dehydroxybaccatin III and 10-deacetylpaclitaxel, semi-synthetic precursors of paclitaxel, from cell cultures of Taxus chinensis, *J. Chromatogr. A 1123*, 15–21.

60. Otieno, D. A., Jondiko, I. J., Mcdowell, P. G., and Kezdy, F. J. (1982) Quantitative-analysis of the pyrethrins by Hplc, *J. Chromatogr. Sci. 20*, 566–570.

61. Hitmi, A., Coudret, A., and Barthomeuf, C. (2000) The production of pyrethrins by plant cell and tissue cultures of Chrysanthemum cinerariaefolium and Tagetes species, *Crit. Rev. Biochem. Mol. Biol. 35*, 317–337.

62. Khanna, P., Sharma, R., and Khanna, R. (1975) Pyrethrins from invivo and invitro tissue-culture of Tagetes-Erecta Linn, *Ind. J. Exper. Biol. 13*, 508–509.

63. Khanna, P., and Khanna, R. (1976) Endogenous free ascorbic-acid and effect of exogenous ascorbic-acid on growth and production of pyrethrins from invitro tissue-culture of Tagetes-Erecta L, *Ind. J. Exper. Biol. 14*, 630–631.

64. Luo, X. D., and Shen, C. C. (1987) The chemistry, pharmacology, and clinical applications of qinghaosu (artemisinin) and its derivatives, *Med. Res. Rev. 7*, 29–52.

65. Liu, D., Zhen, W., Yang, Z., Carter, J. D., Si, H., and Reynolds, K. A. (2006) Genistein acutely stimulates insulin secretion in pancreatic {beta}-cells through a cAMP-dependent protein kinase pathway, *Diabetes 55*, 1043–1050.

66. Rathod, P. K., McErlean, T., and Lee, P. C. (1997) Variations in frequencies of drug resistance in Plasmodium falciparum, *Proc. Natl. Acad. Sci. USA 94*, 9389–9393.

67. Klayman, D. L. (1985) Qinghaosu (artemisinin): An antimalarial drug from China, *Science 228*, 1049–1055.

68. Klayman, D. L., Lin, A. J., Acton, N., Scovill, J. P., Hoch, J. M., Milhous, W. K., Theoharides, A. D., and Dobek, A. S. (1984) Isolation of artemisinin (qinghaosu) from Artemisia annua growing in the United States, *J. Nat. Prod. 47*, 715–717.

69. Nair, M. S., Acton, N., Klayman, D. L., Kendrick, K., Basile, D. V., and Mante, S. (1986) Production of artemisinin in tissue cultures of Artemisia annua, *J. Nat. Prod. 49*, 504–507.

70. Woerdenbag, H. J., Lugt, C. B., and Pras, N. (1990) Artemisia annua L.: A source of novel antimalarial drugs, *Pharm. Weekbl. Sci. 12*, 169–181.

71. Woerdenbag, H. J., Pras, N., van Uden, W., Wallaart, T. E., Beekman, A. C., and Lugt, C. B. (1994) Progress in the research of artemisinin-related antimalarials: An update, *Pharm. World Sci. 16*, 169–180.

72. Schmid, G., and Hofheinz, W. (1983) Total synthesis of qinghaosu, *J. Am. Chem. Soc. 105*, 624–625.

73. Ravindranathan, T., Kumar, M. A., Menon, R. B., and Hiremath, S. V. (1990) Stereoselective synthesis of artemisinin, *Tetrahedr. Lett. 31*, 755–758.

74. Avery, M. A., Chong, W. K. M., and Jenningswhite, C. (1992) Stereoselective total synthesis of (+)-artemisinin, the antimalarial constituent of Artemisia-annua L, *J. Am. Chem. Soc. 114*, 974–979.

75. Haynes, R. K., and Vonwiller, S. C. (1992) Efficient preparation of novel qinghaosu (artemisinin) derivatives—Conversion of qinghao (artemisinic) acid into deoxoqinghaosu derivatives and 5-carba-4-deoxoartesunic acid, *Synlett, 6*, 481–483.

76. Haynes, R. K., and Vonwiller, S. C. (1997) From Qinghao, marvelous herb of antiquity, to the antimalarial trioxane Qinghaosu—And some remarkable new chemistry, *Acc. Chem. Res. 30*, 73–79.

77. Jung, M., Elsohly, H. N., Croom, E. M., Mcphail, A. T., and Mcphail, D. R. (1986) Practical conversion of artemisinic acid into desoxyartemisinin, *J. Org. Chem. 51*, 5417–5419.

78. Roth, R. J., and Acton, N. (1989) A simple conversion of artemisinic acid into artemisinin, *J. Nat. Prod. 52*, 1183–1185.

79. Roth, R. J., and Acton, N. (1987) Isolation of epi-deoxyarteannuin-B from Artemisia-annua, *Planta Med. 53*, 576–576.

80. Roth, R. J., and Acton, N. (1987) Isolation of arteannuic acid from Artemisia-annua, *Planta Med. 53*, 501–502.

81. Dhingra, V., Pakki, S. R., and Narasu, M. L. (2000) Antimicrobial activity of artemisinin and its precursors, *Curr. Sci. 78*, 709–713.

82. Dhingra, V., Rao, K. V., and Narasu, M. L. (1999) Artemisinin: Present status and perspectives, *Biochem. Ed. 27*, 105–109.

83. Dhingra, V., Rao, K. V., and Narasu, M. L. (2000) Current status of artemisinin and its derivatives as antimalarial drugs, *Life Sci. 66*, 279–300.

84. Woerdenbag, H. J., Luers, J. F. J., van Uden, W., Pras, N., Malingre, T. M., and Alfermann, A. W. (1993) Production of the new antimalarial drug artemisinin in shoot cultures of Artemisia-annua L, *Plant Cell Tissue Org. Cult. 32*, 247–257.

85. Liu, B., Ye, H., Li, G., Chen, D., Geng, S., Zhang, Y., Chen, J., and Gao, J. (1998) Studies on dynamics of growth and biosynthesis of artemisinin in hairy roots of Artemisia annua L, *Chin. J. Biotechnol. 14*, 249–254.

86. Liu, C., Wang, Y., Guo, C., Ouyang, F., Ye, H., and Li, G. (1998) Enhanced production of artemisinin by Artemisia annua L hairy root cultures in a modified inner-loop airlift bioreactor, *Bioprocess Eng. 19*, 389–392.

87. Liu, C. Z., Guo, C., Wang, Y. C., and Fan, O. Y. (2003) Factors influencing artemisinin production from shoot cultures of Artemisia annua L., *World J. Microbiol. Biotechnol. 19*, 535–538.

88. Liu, C. Z., Guo, C., Wang, Y. C., and Ouyang, F. (2002) Effect of light irradiation on hairy root growth and artemisinin biosynthesis of Artemisia annua L, *Process Biochem. 38*, 581–585.

89. Liu, C. Z., Wang, Y. C., Guo, C., Ouyang, F., Ye, H. C., and Li, G. F. (1998) Production of artemisinin by shoot cultures of Artemisia annua L. in a modified inner-loop mist bioreactor, *Plant Sci. 135*, 211–217.

90. Xie, D. Y., Zou, Z. R., Ye, H. C., Li, G. F., and Guo, Z. C. (2001) Selection of hairy root clones of Artemisia annua L. for artemisinin production, *Israel J. Plant Sci. 49*, 129–134.

91. Basile, D. V., Akhtari, N., Durand, Y., and Nair, M. S. R. (1993) Toward the production of artemisinin through tissue-culture—Determining nutrient-hormone combinations suitable for cell-suspension cultures, *In Vitro Cell. Dev. Biol.-Plant 29P*, 143–147.

92. Weathers, P. J., Bunk, G., and McCoy, M. C. (2005) The effect of phytohormones on growth and artemisinin production in Artemisia annua hairy roots, *In Vitro Cell. Dev. Biol.-Plant 41*, 47–53.

93. Chen, X. Y., and Xu, Z. H. (1996) Recent progress in biotechnology and genetic engineering of medicinal plants in China, *Med. Chem. Res. 6*, 215–224.

94. Gomez-Galera, S., Pelacho, A. M., Gene, A., Capell, T., and Christou, P. (2007) The genetic manipulation of medicinal and aromatic plants, *Plant Cell Rep. 26*, 1689–1715.

95. Kim, Y., Wyslouzil, B. E., and Weathers, P. J. (2002) Invited review: Secondary metabolism of hairy root cultures in bioreactors, *In Vitro Cell. Dev. Biol.-Plant 38*, 1–10.

96. Chen, D., Ye, H., and Li, G. (2000) Expression of a chimeric farnesyl diphosphate synthase gene in Artemisia annua L. transgenic plants via Agrobacterium tumefaciens-mediated transformation, *Plant Sci. 155*, 179–185.

97. Hezari, M., Lewis, N. G., and Croteau, R. (1995) Purification and characterization of taxa-4(5),11(12)-diene synthase from Pacific yew (Taxus brevifolia) that catalyzes the first committed step of taxol biosynthesis, *Arch. Biochem. Biophys. 322*, 437–444.

98. Martin, V. J., Pitera, D. J., Withers, S. T., Newman, J. D., and Keasling, J. D. (2003) Engineering a mevalonate pathway in *Escherichia coli* for production of terpenoids, *Nat. Biotechnol. 21*, 796–802.

99. Ro, D. K., Paradise, E. M., Ouellet, M., Fisher, K. J., Newman, K. L., Ndungu, J. M., Ho, K. A., Eachus, R. A., Ham, T. S., Kirby, J., Chang, M. C., Withers, S. T., Shiba, Y., Sarpong, R., and Keasling, J. D. (2006) Production of the antimalarial drug precursor artemisinic acid in engineered yeast, *Nature 440*, 940–943.

100. Ro, D. K., Paradise, E. M., Ouellet, M., Fisher, K. J., Newman, K. L., Ndungu, J. M., Ho, K. A., Eachus, R. A., Ham, T. S., Kirby, J., Chang, M. C. Y., Withers, S. T., Shiba, Y., Sarpong, R., and Keasling, J. D. (2006) Production of the antimalarial drug precursor artemisinic acid in engineered yeast, *Nature 440*, 940–943.

101. Lin, X., Hezari, M., Koepp, A. E., Floss, H. G., and Croteau, R. (1996) Mechanism of taxadiene synthase, a diterpene cyclase that catalyzes the first step of taxol biosynthesis in Pacific yew, *Biochemistry 35*, 2968–2977.

102. Huang, Q., Roessner, C. A., Croteau, R., and Scott, A. I. (2001) Engineering Escherichia coli for the synthesis of taxadiene, a key intermediate in the biosynthesis of taxol, *Bioorg. Med. Chem. 9*, 2237–2242.

103. Dejong, J. M., Liu, Y., Bollon, A. P., Long, R. M., Jennewein, S., Williams, D., and Croteau, R. B. (2006) Genetic engineering of taxol biosynthetic genes in Saccharomyces cerevisiae, *Biotechnol. Bioeng. 93*, 212–224.

104. Jennewein, S., Park, H., DeJong, J. M., Long, R. M., Bollon, A. P., and Croteau, R. B. (2005) Coexpression in yeast of Taxus cytochrome P450 reductase with cytochrome P450 oxygenases involved in Taxol biosynthesis, *Biotechnol. Bioeng. 89*, 588–598.

105. Al-Wadei, H. A., Majidi, M., Tsao, M. S., and Schuller, H. M. (2007) Low concentrations of beta-carotene stimulate the proliferation of human pancreatic duct epithelial cells in a PKA-dependent manner, *Cancer Genom. Proteom. 4*, 35–42.

106. Beketova, N. A., Derbeneva, S. A., Spirichev, V. B., Pereverzeva, O. G., Kosheleva, O. V., Mal'tsev, G., Vasil'ev, A. V., and Pogozheva, A. V. (2007) Serum levels of antioxidant and lipid metabolism in patients with cardiovascular disease [in Russian], *Vopr. Pitan. 76*, 11–18.

107. Shih, C. K., Chang, J. H., Yang, S. H., Chou, T. W., and Cheng, H. H. (2008) beta-Carotene and canthaxanthin alter the pro-oxidation and antioxidation balance in rats fed a high-cholesterol and high-fat diet, *Br. J. Nutr. 99*, 1–8.

108. Cui, Y., Lu, Z., Bai, L., Shi, Z., Zhao, W. E., and Zhao, B. (2007) Beta-carotene induces apoptosis and up-regulates peroxisome proliferator-activated receptor gamma expression and reactive oxygen species production in MCF-7 cancer cells, *Eur. J. Cancer. 43*, 2590–2601.

109. Hessel, S., Eichinger, A., Isken, A., Amengual, J., Hunzelmann, S., Hoeller, U., Elste, V., Hunziker, W., Goralczyk, R., Oberhauser, V., von Lintig, J., and Wyss, A. (2007) CMO1-deficiency abolishes vitamin a production from beta-carotene and alters lipid metabolism in mice, *J. Biol. Chem. 282*, 33553–33561.

110. Diretto, G., Al-Babili, S., Tavazza, R., Papacchioli, V., Beyer, P., and Giuliano, G. (2007) Metabolic engineering of potato carotenoid content through tuber-specific overexpression of a bacterial mini-pathway, *PLoS ONE 2*, e350.

111. Wurbs, D., Ruf, S., and Bock, R. (2007) Contained metabolic engineering in tomatoes by expression of carotenoid biosynthesis genes from the plastid genome, *Plant J. 49*, 276–288.

112. Diretto, G., Tavazza, R., Welsch, R., Pizzichini, D., Mourgues, F., Papacchioli, V., Beyer, P., and Giuliano, G. (2006) Metabolic engineering of potato tuber carotenoids through tuber-specific silencing of lycopene epsilon cyclase, *BMC Plant Biol. 6*, 13.

113. al-Babili, S., Ye, X., Lucca, P., Potrykus, I., and Beyer, P. (2001) Biosynthesis of beta-carotene (provitamin A) in rice endosperm achieved by genetic engineering, *Novartis Found. Symp. 236*, 219–228; discussion 228–232.

114. Ye, X., Al-Babili, S., Kloti, A., Zhang, J., Lucca, P., Beyer, P., and Potrykus, I. (2000) Engineering the provitamin A (beta-carotene) biosynthetic pathway into (carotenoid-free) rice endosperm, *Science 287*, 303–305.

115. Beyer, P., Al-Babili, S., Ye, X., Lucca, P., Schaub, P., Welsch, R., and Potrykus, I. (2002) Golden rice: Introducing the beta-carotene biosynthesis pathway into rice endosperm by genetic engineering to defeat vitamin A deficiency, *J. Nutr. 132*, 506S–510S.

116. Lichtenthaler, H. K. (2007) Biosynthesis, accumulation and emission of carotenoids, alpha-tocopherol, plastoquinone, and isoprene in leaves under high photosynthetic irradiance, *Photosynthesis Res. 92*, 163–179.

117. Gerjets, T., Sandmann, M., Zhu, C., and Sandmann, G. (2007) Metabolic engineering of ketocarotenoid biosynthesis in leaves and flowers of tobacco species, *Biotechnol. J. 2*, 1263–1269.

118. Rodriguez-Bustamante, E., and Sanchez, S. (2007) Microbial production of C13-norisoprenoids and other aroma compounds via carotenoid cleavage, *Crit. Rev. Microbiol. 33*, 211–230.

119. Tao, L., Yao, H., and Cheng, Q. (2007) Genes from a Dietzia sp. for synthesis of C40 and C50 beta-cyclic carotenoids, *Gene 386*, 90–97.

120. Sandmann, G., and Misawa, N. (1992) New functional assignment of the carotenogenic genes crtB and crtE with constructs of these genes from Erwinia species, *FEMS Microbiol. Lett. 69*, 253–257.

121. Sandmann, G., Woods, W. S., and Tuveson, R. W. (1990) Identification of carotenoids in Erwinia herbicola and in a transformed Escherichia coli strain, *FEMS Microbiol. Lett. 59*, 77–82.

122. Schnurr, G., Schmidt, A., and Sandmann, G. (1991) Mapping of a carotenogenic gene cluster from Erwinia herbicola and functional identification of six genes, *FEMS Microbiol. Lett. 62*, 157–161.

123. Carmona, M. L., Naganuma, T., and Yamaoka, Y. (2003) Identification by HPLC-MS of carotenoids of the Thraustochytrium CHN-1 strain isolated from the Seto Inland Sea, *Biosci. Biotechnol. Biochem. 67*, 884–888.

124. Di Accadia, F. D., Gribanovski-Sassu, O., Romagnoli, A., and Tuttobello, L. (1966) Isolation and identification of carotenoids produced by a green alga (Dictyococcus cinnabarinus) in submerged culture, *Biochem. J. 101*, 735–740.

125. Giovannucci, D. R., and Stephenson, R. S. (1999) Identification and distribution of dietary precursors of the Drosophila visual pigment chromophore: Analysis of carotenoids in wild type and ninaD mutants by HPLC, *Vision Res. 39*, 219–229.

126. Kushwaha, S. C., and Kates, M. (1973) Isolation and identification of "bacteriorhodopsin" and minor C40-carotenoids in Halobacterium cutirubrum, *Biochim. Biophys. Acta 316*, 235–243.

127. Su, Q., Rowley, K. G., Itsiopoulos, C., and O'Dea, K. (2002) Identification and quantitation of major carotenoids in selected components of the Mediterranean diet: green leafy vegetables, figs and olive oil, *Eur. J. Clin. Nutr. 56*, 1149–1154.

128. Turian, G. (1960) Identification of the major carotenoids of some fungi of the Ascomycetes and Basidiomycetes group. Neurosporene in Cantharellus infundibul informis [in German], *Arch. Mikrobiol. 36*, 139–146.

129. Vanhaelen, M. (1973) Identification of carotenoids in Arnica montana [in French], *Planta Med. 23*, 308–311.

130. Cunningham, F. X., Jr., Pogson, B., Sun, Z., McDonald, K. A., DellaPenna, D., and Gantt, E. (1996) Functional analysis of the beta and epsilon lycopene cyclase enzymes of Arabidopsis reveals a mechanism for control of cyclic carotenoid formation, *Plant Cell 8*, 1613–1626.

131. Yamano, S., Ishii, T., Nakagawa, M., Ikenaga, H., and Misawa, N. (1994) Metabolic engineering for production of beta-carotene and lycopene in Saccharomyces cerevisiae, *Biosci. Biotechnol. Biochem. 58*, 1112–1114.

132. Segre, D., Vitkup, D., and Church, G. M. (2002) Analysis of optimality in natural and perturbed metabolic networks, *Proc. Natl. Acad. Sci. USA 99*, 15112–15117.

133. Varma, A., Boesch, B. W., and Palsson, B. O. (1993) Stoichiometric interpretation of Escherichia coli glucose catabolism under various oxygenation rates, *Appl. Environ. Microbiol. 59*, 2465–2473.

134.   Alexeyev, M. F., and Shokolenko, I. N. (1995) Mini-Tn10 transposon derivatives for insertion mutagenesis and gene delivery into the chromosome of gram-negative bacteria, *Gene 160*, 59–62.

135.   Alper, H., Jin, Y. S., Moxley, J. F., and Stephanopoulos, G. (2005) Identifying gene targets for the metabolic engineering of lycopene biosynthesis in Escherichia coli, *Metab. Eng. 7*, 155–164.

136.   Alper, H., Miyaoku, K., and Stephanopoulos, G. (2005) Construction of lycopene-overproducing E. coli strains by combining systematic and combinatorial gene knockout targets, *Nat. Biotechnol. 23*, 612–616.

137.   Kang, M. J., Lee, Y. M., Yoon, S. H., Kim, J. H., Ock, S. W., Jung, K. H., Shin, Y. C., Keasling, J. D., and Kim, S. W. (2005) Identification of genes affecting lycopene accumulation in Escherichia coli using a shot-gun method, *Biotechnol. Bioeng. 91*, 636–642.

138.   Becker-Hapak, M., Troxtel, E., Hoerter, J., and Eisenstark, A. (1997) RpoS dependent overexpression of carotenoids from Erwinia herbicola in OXYR deficient Escherichia coli, *Biochem. Biophys. Res. Commun. 239*, 305–309.

139.   Sun, Z., Cunningham, F. X., Jr., and Gantt, E. (1998) Differential expression of two isopentenyl pyrophosphate isomerases and enhanced carotenoid accumulation in a unicellular chlorophyte, *Proc. Natl. Acad. Sci. USA 95*, 11482–11488.

140.   Neuman, I., Nahum, H., and Ben-Amotz, A. (2000) Reduction of exercise-induced asthma oxidative stress by lycopene, a natural antioxidant, *Allergy 55*, 1184–1189.

141.   Nishino, H., Murakosh, M., Ii, T., Takemura, M., Kuchide, M., Kanazawa, M., Mou, X. Y., Wada, S., Masuda, M., Ohsaka, Y., Yogosawa, S., Satomi, Y., and Jinno, K. (2002) Carotenoids in cancer chemoprevention, *Cancer Metastasis Rev. 21*, 257–264.

142.   Hix, L. M., Lockwood, S. F., and Bertram, J. S. (2004) Bioactive carotenoids: Potent antioxidants and regulators of gene expression, *Redox Rep. 9*, 181–191.

143.   Lee, J. H., and Kim, Y. T. (2006) Cloning and characterization of the astaxanthin biosynthesis gene cluster from the marine bacterium Paracoccus haeundaensis, *Gene 370*, 86–95.

144.   Lee, J. H., and Kim, Y. T. (2006) Functional expression of the astaxanthin biosynthesis genes from a marine bacterium, Paracoccus haeundaensis, *Biotechnol. Lett. 28*, 1167–1173.

145.   Tao, L., Wilczek, J., Odom, J. M., and Cheng, Q. (2006) Engineering a beta-carotene ketolase for astaxanthin production, *Metab. Eng. 8*, 523–531.

146.   Farmer, W. R., and Liao, J. C. (2001) Precursor balancing for metabolic engineering of lycopene production in Escherichia coli, *Biotechnol. Prog. 17*, 57–61.

147.   Tao, L., Sedkova, N., Yao, H., Ye, R. W., Sharpe, P. L., and Cheng, Q. (2007) Expression of bacterial hemoglobin genes to improve astaxanthin production in a methanotrophic bacterium Methylomonas sp, *Appl. Microbiol. Biotechnol. 74*, 625–633.

148.   Del Rio, E., Acien, F. G., Garcia-Malea, M. C., Rivas, J., Molina-Grima, E., and Guerrero, M. G. (2005) Efficient one-step production of astaxanthin by the microalga Haematococcus pluvialis in continuous culture, *Biotechnol. Bioeng. 91*, 808–815.

149. Yuan, L. Z., Rouviere, P. E., Larossa, R. A., and Suh, W. (2006) Chromosomal promoter replacement of the isoprenoid pathway for enhancing carotenoid production in E. coli, *Metab. Eng. 8*, 79–90.

150. Sacchettini, J. C., and Poulter, C. D. (1997) Creating isoprenoid diversity, *Science 277*, 1788–1789.

151. Tobias, A. V., and Arnold, F. H. (2006) Biosynthesis of novel carotenoid families based on unnatural carbon backbones: A model for diversification of natural product pathways, *Biochim. Biophys. Acta 1761*, 235–246.

152. Ruther, A., Misawa, N., Boger, P., and Sandmann, G. (1997) Production of zeaxanthin in Escherichia coli transformed with different carotenogenic plasmids, *Appl. Microbiol. Biotechnol. 48*, 162–167.

153. Schmidt-Dannert, C., Umeno, D., and Arnold, F. H. (2000) Molecular breeding of carotenoid biosynthetic pathways, *Nat. Biotechnol. 18*, 750–753.

154. Tao, L., Jackson, R. E., Rouviere, P. E., and Cheng, Q. (2005) Isolation of chromosomal mutations that affect carotenoid production in Escherichia coli: Mutations alter copy number of ColE1-type plasmids, *FEMS Microbiol. Lett. 243*, 227–233.

155. Negishi, E., Liou, S. Y., Xu, C., and Huo, S. (2002) A novel, highly selective, and general methodology for the synthesis of 1,5-diene-containing oligoisoprenoids of all possible geometrical combinations exemplified by an iterative and convergent synthesis of coenzyme Q(10), *Org. Lett. 4*, 261–264.

156. Lipshutz, B. H., Mollard, P., Pfeiffer, S. S., and Chrisman, W. (2002) A short, highly efficient synthesis of coenzyme Q(10), *J. Am. Chem. Soc. 124*, 14282–14283.

157. Park, Y. C., Kim, S. J., Choi, J. H., Lee, W. H., Park, K. M., Kawamukai, M., Ryu, Y. W., and Seo, J. H. (2005) Batch and fed-batch production of coenzyme Q10 in recombinant Escherichia coli containing the decaprenyl diphosphate synthase gene from Gluconobacter suboxydans, *Appl. Microbiol. Biotechnol. 67*, 192–196.

158. Yoshida, H., Kotani, Y., Ochiai, K., and Araki, K. (1998) Production of ubiquinone-10 using bacteria, *J. Gen. Appl. Microbiol. 44*, 19–26.

159. Peterson, J. K., Tucker, C., Favours, E., Cheshire, P. J., Creech, J., Billups, C. A., Smykla, R., Lee, F. Y., and Houghton, P. J. (2005) In vivo evaluation of ixabepilone (BMS247550), a novel epothilone B derivative, against pediatric cancer models, *Clin. Cancer. Res. 11*, 6950–6958.

160. Frykman, S. A., Tsuruta, H., and Licari, P. J. (2005) Assessment of fed-batch, semi-continuous, and continuous epothilone D production processes, *Biotechnol. Prog. 21*, 1102–1108.

161. Jumaa, M., Carlson, B., Chimilio, L., Silchenko, S., and Stella, V. J. (2004) Kinetics and mechanism of degradation of epothilone-D: An experimental anticancer agent, *J. Pharm. Sci. 93*, 2953–2961.

162. Plunkett, J. W. (2006) Plunkett's Biotech & Genetics Industry Almanac 2007: Biotech & Genetics Industry Market Research, Statistics, Trends & Leading Companies, Plunkett Research, Houston, TX, p. 578.

163. Woodward, R. B., Logusch, E., Nambiar, K. P., Sakan, K., Ward, D. E., Au-Yeung, B. W., Balaram, P., Browne, L. J., Card, P. J., and Chen, C. H. (1981) Asymmetric total synthesis of erythromcin. 1. Synthesis of an erythronolide A secoacid derivative via asymmetric induction, *J. Am. Chem. Soc. 103*, 3210–3213.

164. Masamune, S., Hirama, M., Mori, S., Ali, S. A., and Garvey, D. S. (1981) Total synthesis of 6-deoxyerythronolide B, *J. Am. Chem. Soc. 103*, 1568–1571.

165. Smith, A. B., 3rd, Walsh, S. P., Frohn, M., and Duffey, M. O. 2005 Diversity-oriented synthesis of polyketide natural products via iterative chemo- and stereoselective functionalization of polyenoates: Development of a unified approach for the C(1–19) segments of lituarines A–C, *Org. Lett. 7* 139–142.

166. Cheng, Q., Xiang, L., Izumikawa, M., Meluzzi, D., and Moore, B. S. (2007) Enzymatic total synthesis of enterocin polyketides, *Nat. Chem. Biol. 3*, 557–558.

167. Tatsuta, K., and Hosokawa, S. (2006) Total syntheses of polyketide-derived bioactive natural products, *Chem. Rec. 6*, 217–233.

168. Enomoto, M., and Kuwahara, S. (2006) Enantioselective total synthesis and stereochemical revision of communiols E and F, *J. Org. Chem. 71*, 6287–6290.

169. Schetter, B., and Mahrwald, R. (2006) Modern aldol methods for the total synthesis of polyketides, *Ang. Chem. Int. ed. 45*, 7506–7525.

170. Woodward, R. B. (1956) Perspecitves in Organic Chemistry, Wiley-Interscience, New York.

171. Austin, M. B., and Noel, J. P. (2003) The chalcone synthase superfamily of type III polyketide synthases, *Nat. Prod. Rep. 20*, 79–110.

172. Abe, I., Sano, Y., Takahashi, Y., and Noguchi, H. (2003) Site-directed mutagenesis of benzalacetone synthase. The role of the Phe215 in plant type III polyketide synthases, *J. Biol. Chem. 278*, 25218–25226.

173. Su, D.-S., Meng, D. F., Bertinato, P., Balog, A., Sorensen, E. J., Danishefsky, S. J., Zheng, Y.-H., Chou, T.-C., He, L.F., and Horwitz, S. B. (1997) Total synthesis of (−)- epothilone B: An extension of the Suzuki coupling method and insights into structure-activity relationships of the epothilones, *Ang. Chem. Int. Ed. Engl. 36*, 757–759.

174. Bode, J. W., and Carreira, E. M. (2001) Stereoselective syntheses of epothilones A and B via directed nitrile oxide cycloaddition, *J. Am. Chem. Soc. 123*, 3611–3612.

175. Mulzer, J., Mantoulidis, A., and Ohler, E. (2000) Total syntheses of epothilones B and D, *J. Org. Chem. 65*, 7456–7467.

176. Hearn, B. R., Zhang, D., Li, Y., and Myles, D. C. (2006) C-15 thiazol-4-yl analogues of (*E*)-9,10-didehydroepothilone D: Synthesis and cytotoxicity, *Org. Lett. 8*, 3057–3059.

177. Alhamadsheh, M. M., Hudson, R. A., and Viranga Tillekeratne, L. M. (2006) Design, total synthesis, and evaluation of novel open-chain epothilone analogues, *Org. Lett. 8*, 685– 688.

178. Boddy, C. N., Hotta, K., Tse, M. L., Watts, R. E., and Khosla, C. (2004) Precursor-directed biosynthesis of epothilone in Escherichia coli, *J. Am. Chem. Soc. 126*, 7436–7437.

179. Starks, C. M., Zhou, Y., Liu, F., and Licari, P. J. (2003) Isolation and characterization of new epothilone analogues from recombinant Myxococcus xanthus fermentations, *J. Nat. Prod. 66*, 1313–1317.

180. Kealey, J. T. (2003) Creating polyketide diversity through genetic engineering, *Front. Biosci. 8*, c1–13.

181. Pfeifer, B. A., Admiraal, S. J., Gramajo, H., Cane, D. E., and Khosla, C. (2001) Biosynthesis of complex polyketides in a metabolically engineered strain of E. coli, *Science 291*, 1790–1792.

182. Julien, B., Shah, S., Ziermann, R., Goldman, R., Katz, L., and Khosla, C. (2000) Isolation and characterization of the epothilone biosynthetic gene cluster from Sorangium cellulosum, *Gene 249*, 153–160.

183. Gerth, K., Steinmetz, H., Hofle, G., and Reichenbach, H. (2000) Studies on the biosynthesis of epothilones: The biosynthetic origin of the carbon skeleton, *J. Antibiot. (Tokyo) 53*, 1373–1377.

184. Gerth, K., Steinmetz, H., Hofle, G., and Reichenbach, H. (2001) Studies on the biosynthesis of epothilones: The PKS and Epothilone C/D monooxygenase, *J. Antibiot. (Tokyo) 54*, 144–148.

185. Tang, L., Shah, S., Chung, L., Carney, J., Katz, L., Khosla, C., and Julien, B. (2000) Cloning and heterologous expression of the epothilone gene cluster, *Science 287*, 640–642.

186. Julien, B., and Shah, S. (2002) Heterologous expression of epothilone biosynthetic genes in Myxococcus xanthus, *Antimicrob. Agents Chemother. 46*, 2772–2778.

187. Lau, J., Frykman, S., Regentin, R., Ou, S., Tsuruta, H., and Licari, P. (2002) Optimizing the heterologous production of epothilone D in Myxococcus xanthus, *Biotechnol. Bioeng. 78*, 280–288.

188. Mutka, S. C., Carney, J. R., Liu, Y., and Kennedy, J. (2006) Heterologous production of epothilone C and D in Escherichia coli, *Biochemistry 45*, 1321–1330.

189. Donadio, S., and Katz, L. (1992) Organization of the enzymatic domains in the multifunctional polyketide synthase involved in erythromycin formation in Saccharopolyspora erythraea, *Gene 111*, 51–60.

190. Kao, C. M., Katz, L., and Khosla, C. (1994) Engineered biosynthesis of a complete macrolactone in a heterologous host, *Science 265*, 509–512.

191. Bruheim, P., Sletta, H., Bibb, M. J., White, J., and Levine, D. W. (2002) High-yield actinorhodin production in fed-batch culture by a Streptomyces lividans strain overexpressing the pathway-specific activator gene actll-ORF4, *J. Ind. Microbiol. Biotechnol. 28*, 103–111.

192. Donadio, S., McAlpine, J. B., Sheldon, P. J., Jackson, M., and Katz, L. (1993) An erythromycin analog produced by reprogramming of polyketide synthesis, *Proc. Natl. Acad. Sci. USA 90*, 7119–7123.

193. Cortes, J., Wiesmann, K. E., Roberts, G. A., Brown, M. J., Staunton, J., and Leadlay, P. F. (1995) Repositioning of a domain in a modular polyketide synthase to promote specific chain cleavage, *Science 268*, 1487–1489.

194. Stassi, D. L., Kakavas, S. J., Reynolds, K. A., Gunawardana, G., Swanson, S., Zeidner, D., Jackson, M., Liu, H., Buko, A., and Katz, L. (1998) Ethyl-substituted erythromycin derivatives produced by directed metabolic engineering, *Proc. Natl. Acad. Sci. USA 95*, 7305–7309.

195. Jacobsen, J. R., Hutchinson, C. R., Cane, D. E., and Khosla, C. (1997) Precursor-directed biosynthesis of erythromycin analogs by an engineered polyketide synthase, *Science 277*, 367–369.

196. Zhang, Y. X., Perry, K., Vinci, V. A., Powell, K., Stemmer, W. P., and del Cardayre, S. B. (2002) Genome shuffling leads to rapid phenotypic improvement in bacteria, *Nature 415*, 644–646.

197. Liutskanova, D. G., Stoilova-Disheva, M. M., and Peltekova, V. T. (2005) Increase in tylosin production by a commercial strain of Streptomyces fradiae [in Russian], *Prikl. Biokhim. Mikrobiol. 41*, 189–193.

198. Jung, W. S., Lee, S. K., Hong, J. S., Park, S. R., Jeong, S. J., Han, A. R., Sohng, J. K., Kim, B. G., Choi, C. Y., Sherman, D. H., and Yoon, Y. J. (2006) Heterologous expression of tylosin polyketide synthase and production of a hybrid bioactive macrolide in Streptomyces venezuelae, *Appl. Microbiol. Biotechnol. 72*, 763–769.

199. Bate, N., Bignell, D. R., and Cundliffe, E. (2006) Regulation of tylosin biosynthesis involving "SARP-helper" activity, *Mol. Microbiol. 62*, 148–156.

200. Bignell, D. R., Bate, N., and Cundliffe, E. (2007) Regulation of tylosin production: Role of a TylP-interactive ligand, *Mol. Microbial. 63*, 838–847.

201. Li, K., Mikola, M. R., Draths, K. M., Worden, R. M., and Frost, J. W. (1999) Fed-batch fermentor synthesis of 3-dehydroshikimic acid using recombinant Escherichia coli, *Biotechnol. Bioeng. 64*, 61–73.

202. Harborne, J. B., and Williams, C. A. (2000) Advances in flavonoid research since 1992, *Phytochemistry 55*, 481–504.

203. Block, G., Patterson, B., and Subar, A. (1992) Fruit, vegetables, and cancer prevention: A review of the epidemiological evidence, *Nutr. Cancer 18*, 1–29.

204. Birt, D. F., Hendrich, S., and Wang, W. (2001) Dietary agents in cancer prevention: Flavonoids and isoflavonoids, *Pharmacol. Ther. 90*, 157–177.

205. Kobayashi, Y., Akita, M., Sakamoto, K., Liu, H., Shigeoka, T., Koyano, T., Kawamura, M., and Furuya, T. (1993) Large-scale production of anthocyanin by *Aralia cordata* cell suspension cultures, *Appl. Microbiol. Biotechnol. 40*, 215–218.

206. Zhong, J.-J., Seki, T., Kinoshita, S.-I., and Yoshida, T. (1991) Effect of light irradiation on anthocyanin production by suspended culture of *Perilla frutescens, Biotechnol. Bioeng. 38*, 653–658.

207. Liu, C. J., Blount, J. W., Steele, C. L., and Dixon, R. A. (2002) Bottlenecks for metabolic engineering of isoflavone glycoconjugates in *Arabidopsis, Proc. Natl. Acad. Sci. USA 99*, 14578–14583.

208. Yu, O., Shi, J., Hession, A. O., Maxwell, C. A., McGonigle, B., and Odell, J. T. (2003) Metabolic engineering to increase isoflavone biosynthesis in soybean seed, *Phytochemistry 63*, 753–763.

209. Wang, H. J., and Murphy, P. A. (1996) Mass balance study of isoflavones during soybean processing, *J. Agric. Food Chem. 44*, 2377–2383.

210. Furlong, J. J. P., and Nudelman, N. S. (1985) Mechanism of cyclization of substituted 2′-hydroxychalcones to flavanones, *J. Chem. Soc. Perk. T 2*, 633–639.

211. Lim, S. S., Jung, S. H., Ji, J., Shin, K. H., and Keum, S. R. (2001) Synthesis of flavonoids and their effects on aldose reductase and sorbitol accumulation in streptozotocin-induced diabetic rat tissues, *J. Pharm. Pharmacol. 53*, 653–668.

212. Kumazawa, T., Kimura, T., Matsuba, S., Sato, S., and Onodera, J. (2001) Synthesis of 8-*C*-glucosylflavones, *Carbohyd. Res. 334*, 183–193.

213. Tanaka, H., Stohlmeyer, M. M., Wandless, T. J., and Taylor, L. P. (2000) Synthesis of flavonol derivatives as probes of biological processes, *Tetrahedr. Lett. 41*, 9735–9739.

214. Wan, S. B., and Chan, T. H. (2004) Enantioselective synthesis of afzelechin and epiafzelechin, *Tetrahedron 60*, 8207–8211.

215. Onda, M., Li, S. S., Li, X. A., Harigaya, Y., Takahashi, H., Kawase, H., and Kagawa, H. (1989) Heterocycles .24. Synthesis of optically pure 2,3-trans-5,7,3′,4′,5′-pentahydroxyflavan-3,4-diols and comparison with naturally-occurring leucodelphinidins, *J. Nat. Prod. 52*, 1100–1106.

216. Farina, A., Ferranti, C., and Marra, C. (2006) An improved synthesis of resveratrol, *Nat. Prod. Res. 20*, 247–252.

217. Li, Y. Q., Li, Z. L., Zhao, W. J., Wen, R. X., Meng, Q. W., and Zeng, Y. (2006) Synthesis of stilbene derivatives with inhibition of SARS coronavirus replication, *Eur. J. Med. Chem. 41*, 1084–1089.

218. Takaya, Y., Terashima, K., Ito, J., He, Y. H., Tateoka, M., Yamaguchi, N., and Niwa, M. (2005) Biomimic transformation of resveratrol, *Tetrahedron 61*, 10285–10290.

219. Hwang, E. I., Kaneko, M., Ohnishi, Y., and Horinouchi, S. (2003) Production of plant-specific flavanones by *Escherichia coli* containing an artificial gene cluster, *Appl. Environ. Microbiol. 69*, 2699–2706.

220. Watts, K. T., Lee, P. C., and Schmidt-Dannert, C. (2004) Exploring recombinant flavonoid biosynthesis in metabolically engineered *Escherichia coli*, *Chembiochem 5*, 500–507.

221. Leonard, E., Lim, K. H., Saw, P. N., and Koffas, M. A. (2007) Engineering central metabolic pathways for high-level flavonoid production in Escherichia coli, *Appl. Environ. Microbiol. 73*, 3877–3886.

222. Dixon, R. A., and Sumner, L. W. (2003) Legume natural products: Understanding and manipulating complex pathways for human and animal health, *Plant Physiol. 131*, 878–885.

223. Jung, W., Yu, O., Lau, S. M., O'Keefe, D. P., Odell, J., Fader, G., and McGonigle, B. (2000) Identification and expression of isoflavone synthase, the key enzyme for biosynthesis of isoflavones in legumes, *Nat. Biotechnol. 18*, 208–212.

224. Ralston, L., Subramanian, S., Matsuno, M., and Yu, O. (2005) Partial reconstruction of flavonoid and isoflavonoid biosynthesis in yeast using soybean type I and type II chalcone isomerases, *Plant Physiol. 137*, 1375–1388.

225. Leonard, E., Yan, Y., Lim, K. H., and Koffas, M. A. (2005) Investigation of two distinct flavone synthases for plant-specific flavone biosynthesis in Saccharomyces cerevisiae, *Appl. Environ. Microbiol. 71*, 8241–8248.

226. Leonard, E., and Koffas, M. (2007) Engineering artificial plant cytochrome P450s for isoflavonoid synthesis from Escherichia coli, *Appl. Environ. Microbiol. 73*, 7246–7251.

227. Way, T. D., Kao, M. C., and Lin, J. K. (2004) Apigenin induces apoptosis through proteasomal degradation of HER2/neu in HER2/neu-overexpressing breast cancer cells via the phosphatidylinositol 3-kinase/Akt-dependent pathway, *J. Biol. Chem. 279*, 4479–4489.

228. Caltagirone, S., Rossi, C., Poggi, A., Ranelletti, F. O., Natali, P. G., Brunetti, M., Aiello, F. B., and Piantelli, M. (2000) Flavonoids apigenin and quercetin inhibit melanoma growth and metastatic potential, *Int. J. Cancer 87*, 595–600.

229. Critchfield, J. W., Coligan, J. E., Folks, T. M., and Butera, S. T. (1997) Casein kinase II is a selective target of HIV-1 transcriptional inhibitors, *Proc. Nat. Acad. Sci. USA 94*, 6110–6115.

230. Martens, S., and Forkmann, G. (1999) Cloning and expression of flavone synthase II from Gerbera hybrids, *Plant J. 20*, 611–618.

231. Yan, Y. J., Kohli, A., and Koffas, M. A. G. (2005) Biosynthesis of natural flavanones in Saccharomyces cerevisiae, *Appl. Environ. Microbiol. 71*, 5610–5613.

232. Formica, J. V., and Regelson, W. (1995) Review of the biology of quercetin and related bioflavonoids, *Food Chem. Toxicol. 33*, 1061–1080.

233. Lamson, D. W., and Brignall, M. S. (2000) Antioxidants and cancer, part 3: Quercetin, *Altern. Med. Rev. 5*, 196–208.

234. Leonard, E., Chemler, J., Lim, K. H., and Koffas, M. A. (2006) Expression of a soluble flavone synthase allows the biosynthesis of phytoestrogen derivatives in Escherichia coli, *Appl. Microbiol. Biotechnol. 70*, 85–91.

235. Mol, J., Grotewold, E., and Koes, R. (1998) How genes paint flowers and seeds, *Trends Plant Sci. 3*, 212–217.

236. Weisel, T., Baum, M., Eisenbrand, G., Dietrich, H., Will, F., Stockis, J. P., Kulling, S., Rufer, C., Johannes, C., and Janzowski, C. (2006) An anthocyanin/polyphenolic-rich fruit juice reduces oxidative DNA damage and increases glutathione level in healthy probands, *Biotechnol. J. 1*, 388–397.

237. Lala, G., Malik, M., Zhao, C., He, J., Kwon, Y., Giusti, M. M., and Magnuson, B. A. (2006) Anthocyanin-rich extracts inhibit multiple biomarkers of colon cancer in rats, *Nutr. Cancer 54*, 84–93.

238. Wrolstad, R. E. (2000) Anthocyanins, in *Natural Food Colorants*, Francis, F. J., and Lauro, G. J., Eds., Marcel Dekker, New York, pp. 237–252.

239. Kamei, H., Kojima, T., Hasegawa, M., Koide, T., Umeda, T., Yukawa, T., and Terabe, K. (1995) Suppression of tumor cell growth by anthocyanins in vitro, *Cancer Investig. 13*, 590–594.

240. Koide, T., Kamei, H., Hashimoto, Y., Kojima, T., and Hasegawa, M. (1996) Antitumor effect of hydrolyzed anthocyanin from grape rinds and red rice, *Cancer Biother. Radio. 11*, 273–277.

241. Koide, T., Hashimoto, Y., Kamei, H., Kojima, T., Hasegawa, M., and Terabe, K. (1997) Antitumor effect of anthocyanin fractions extracted from red soybeans and red beans in vitro and in vivo, *Cancer Biother. Radiopharm. 12*, 277–280.

242. Bomser, J., Madhavi, D. L., Singletary, K., and Smith, M. A. (1996) In vitro anticancer activity of fruit extracts from *Vaccinium* species, *Planta Med. 62*, 212–216.

243. Hagiwara, A., Miyashita, K., Nakanishi, T., Sano, M., Tamano, S., Kadota, T., Koda, T., Nakamura, M., Imaida, K., Ito, N., and Shirai, T. (2001) Pronounced inhibition by a natural anthocyanin, purple corn color, of 2-amino-1-methyl-6-phenylimidazo[4,5-*b*] pyridine (PhIP)-associated colorectal carcinogenesis in male F344 rats pretreated with 1,2-dimethylhydrazine, *Cancer Lett. 171*, 17–25.

244. Harris, G. K., Gupta, A., Nines, R. G., Kresty, L. A., Habib, S. G., Frankel, W. L., LaPerle, K., Gallaher, D. D., Schwartz, S. J., and Stoner, G. D. (2001) Effects of lyophilized black

raspberries on azoxymethane-induced colon cancer and 8-hydroxy-2′-deoxyguanosine levels in the Fischer 344 rat, *Nutr. Cancer 40*, 125–133.

245. Kang, S. Y., Seeram, N. P., Nair, M. G., and Bourquin, L. D. (2003) Tart cherry anthocyanins inhibit tumor development in Apc(Min) mice and reduce proliferation of human colon cancer cells, *Cancer Lett. 194*, 13–19.

246. Katsube, N., Iwashita, K., Tsushida, T., Yamaki, K., and Kobori, M. (2003) Induction of apoptosis in cancer cells by Bilberry (*Vaccinium myrtillus*) and the anthocyanins, *J. Agric. Food Chem. 51*, 68–75.

247. Mclellan, M. R., Kime, R. W., Lee, C. Y., and Long, T. M. (1995) Effect of honey as an antibrowning agent in light raisin processing, *J. Food Process. Preserv. 19*, 1–8.

248. Oszmianski, J., and Lee, C. Y. (1990) Inhibition of polyphenol oxidase activity and browning by honey, *J. Agric. Food Chem. 38*, 1892–1895.

249. Tsai, P. J., Hsieh, Y. Y., and Huang, T. C. (2004) Effect of sugar on anthocyanin degradation and water mobility in a roselle anthocyanin model system using 17O NMR, *J. Agric. Food Chem. 52*, 3097–3099.

250. Hanagata, N., Ito, A., Uehara, H., Asari, F., Takeuchi, T., and Karube, I. (1993) Behavior of cell aggregate of *Carthamus tinctorius* L. cultured cells and correlation with red pigment formation, *J. Biotechnol. 30*, 259–269.

251. Wellmann, E. (1975) Uv dose-dependent induction of enzymes related to flavonoid biosynthesis in cell-suspension cultures of parsley, *FEBS Lett. 51*, 105–107.

252. Meyer, J. E., Pepin, M. F., and Smith, M. A. (2002) Anthocyanin production from *Vaccinium pahalae*: Limitations of the physical microenvironment, *J. Biotechnol. 93*, 45–57.

253. Hall, R. D., and Yeoman, M. M. (1986) Temporal and spatial heterogeneity in the accumulation of anthocyanins in cell-cultures of Catharanthus-roseus (L) Don, G, *J. Exp. Botan. 37*, 48–60.

254. Smith, M. A. L., and Spomer, L. A. (1995) Vessels, gels, liquid media, and support systems, in *Automation and Environmental Control in Plant Tissue Culture*, Aitken-Christie, J., Kozai, T., and Smith, M. A. L., Eds., Springer, New York, pp. 371–404.

255. Hellwig, S., Drossard, J., Twyman, R. M., and Fischer, R. (2004) Plant cell cultures for the production of recombinant proteins, *Nat. Biotechnol. 22*, 1415–1422.

256. Yan, Y. J., Chemler, J., Huang, L. X., Martens, S., and Koffas, M. A. G. (2005) Metabolic engineering of anthocyanin biosynthesis in Escherichia coli, *Appl. Environ. Microbiol. 71*, 3617–3623.

257. Howitz, K. T., Bitterman, K. J., Cohen, H. Y., Lamming, D. W., Lavu, S., Wood, J. G., Zipkin, R. E., Chung, P., Kisielewski, A., Zhang, L. L., Scherer, B., and Sinclair, D. A. (2003) Small molecule activators of sirtuins extend Saccharomyces cerevisiae lifespan, *Nature 425*, 191–196.

258. Wood, J. G., Rogina, B., Lavu, S., Howitz, K., Helfand, S. L., Tatar, M., and Sinclair, D. (2004) Sirtuin activators mimic caloric restriction and delay ageing in metazoans, *Nature 430*, 686–689.

259. Viswanathan, M., Kim, S. K., Berdichevsky, A., and Guarente, L. (2005) A role for SIR-2.1 regulation of ER stress response genes in determining C. elegans life span, *Dev. Cell 9*, 605–615.

260. Baur, J. A., Pearson, K. J., Price, N. L., Jamieson, H. A., Lerin, C., Kalra, A., Prabhu, V. V., Allard, J. S., Lopez-Lluch, G., Lewis, K., Pistell, P. J., Poosala, S., Becker, K. G., Boss, O., Gwinn, D., Wang, M., Ramaswamy, S., Fishbein, K. W., Spencer, R. G., Lakatta, E. G., Le Couteur, D., Shaw, R. J., Navas, P., Puigserver, P., Ingram, D. K., de Cabo, R., and Sinclair, D. A. (2006) Resveratrol improves health and survival of mice on a high-calorie diet, *Nature 444*, 337–342.

261. Becker, J. V. W., Armstrong, G. O., Van der Merwe, M. J., Lambrechts, M. G., Vivier, M. A., and Pretorius, I. S. (2003) Metabolic engineering of *Saccharomyces cerevisiae* for the synthesis of the wine-related antioxidant resveratrol, *FEMS Yeast Res. 4*, 79–85.

262. Watts, K. T., Lee, P. C., and Schmidt-Dannert, C. (2006) Biosynthesis of plant-specific stilbene polyketides in metabolically engineered Escherichia coli, *BMC Biotechnol. 6*, article 22.

263. Beekwilder, J., Wolswinkel, R., Jonker, H., Hall, R., de Vos, C. H., and Bovy, A. (2006) Production of resveratrol in recombinant microorganisms, *Appl. Environ. Microbiol. 72*, 5670–5672.

264. Sakaguchi, M., Mihara, K., and Sato, R. (1984) Signal recognition particle is required for co-translational insertion of cytochrome P-450 into microsomal membranes, *Proc. Natl. Acad. Sci. USA 81*, 3361–3364.

265. Williams, P. A., Cosme, J., Sridhar, V., Johnson, E. F., and McRee, D. E. (2000) Microsomal cytochrome P450 2C5: Comparison to microbial P450s and unique features, *J. Inorg. Biochem. 81*, 183–190.

266. Yamazaki, S., Sato, K., Suhara, K., Sakaguchi, M., Mihara, K., and Omura, T. (1993) Importance of the proline-rich region following signal-anchor sequence in the formation of correct conformation of microsomal cytochrome P-450s, *J. Biochem. (Tokyo) 114*, 652–657.

267. Williams, P. A., Cosme, J., Sridhar, V., Johnson, E. F., and McRee, D. E. (2000) Mammalian microsomal cytochrome P450 monooxygenase: Structural adaptations for membrane binding and functional diversity, *Mol. Cell 5*, 121–131.

268. Williams, P. A., Cosme, J., Ward, A., Angove, H. C., Matak Vinkovic, D., and Jhoti, H. (2003) Crystal structure of human cytochrome P450 2C9 with bound warfarin, *Nature 424*, 464–468.

269. Williams, P. A., Cosme, J., Vinkovic, D. M., Ward, A., Angove, H. C., Day, P. J., Vonrhein, C., Tickle, I. J., and Jhoti, H. (2004) Crystal structures of human cytochrome P450 3A4 bound to metyrapone and progesterone, *Science 305*, 683–686.

270. Haudenschild, C., Schalk, M., Karp, F., and Croteau, R. (2000) Functional expression of regiospecific cytochrome P450 limonene hydroxylases from mint (Mentha spp.) in Escherichia coli and saccharomyces cerevisiae, *Arch. Biochem. Biophys. 379*, 127–136.

271. Hotze, M., Schroder, G., and Schroder, J. (1995) Cinnamate 4-hydroxylase from *Catharanthus roseus*, and a strategy for the functional expression of plant cytochrome P450 proteins as translational fusions with P450 reductase in *Escherichia coli*, *FEBS Lett. 374*, 345–350.

272. Looman, A. C., Bodlaender, J., Comstock, L. J., Eaton, D., Jhurani, P., de Boer, H. A., and van Knippenberg, P. H. (1987) Influence of the codon following the AUG initiation codon on the expression of a modified lacZ gene in Escherichia coli, *EMBO J. 6*, 2489–2492.

273. Barnes, H. J., Arlotto, M. P., and Waterman, M. R. (1991) Expression and enzymatic activity of recombinant cytochrome P450 17 alpha-hydroxylase in Escherichia coli, *Proc. Natl. Acad. Sci. USA 88*, 5597–5601.

274. Stormo, G. D., Schneider, T. D., and Gold, L. M. (1982) Characterization of translational initiation sites in E. coli, *Nucl. Acids Res. 10*, 2971–2996.

275. Schauder, B., and McCarthy, J. E. (1989) The role of bases upstream of the Shine-Dalgarno region and in the coding sequence in the control of gene expression in Escherichia coli: Translation and stability of mRNAs in vivo, *Gene 78*, 59–72.

276. Leonard, E., Yan, Y., and Koffas, M. A. (2006) Functional expression of a P450 flavonoid hydroxylase for the biosynthesis of plant-specific hydroxylated flavonols in Escherichia coli, *Metab. Eng. 8*, 172–181.

277. Fox, A. K., Powers, W. M., Rivera, S. A., Arce, H., Deese, W., Grossman, N., Johnson, K., Signoret, J., Tsigas, M., Boron, A., Coleman, J., Corey, R., Freund, K., Gehrke, B., Greenblatt, J., Herfindahl, E., Newman, D., Peterson, J., Serletis, G., Dixon, P., Rimmer, M., Winston, A. (2007) The Economic Effects of Significant U.S. Import Restraints, Fifth Update, USIT Commission, www.usitc.gov/publications/abstract_3906.htm, accessed June 11, 2008.

278. (2006) Annual and Monthly U.S. Ethanol Production, *Biofuels J.*, http://www.grainnet. com/articles/Annual_and_Monthly_U_S__Ethanol_Production-25474.html, accessed June 11, 2008.

279. Feldmann, S. D., Sahm, H., and Sprenger, G. A. (1992) Cloning and expression of the genes for xylose isomerase and xylulokinase from Klebsiella pneumoniae 1033 in Escherichia coli K12, *Mol. Gen. Genet. 234*, 201–210.

280. Hahn-Hägerdal, B., Karhumaa, K., Fonseca, C., Spencer-Martins, I., and Gorwa-Grauslund, M. (2007) Towards industrial pentose-fermenting yeast strains, *Appl. Microbiol. Biotechnol. 74*, 937–953.

281. Alterthum, F., and Ingram, L. O. (1989) Efficient ethanol production from glucose, lactose, and xylose by recombinant Escherichia coli, *Appl. Environ. Microbiol. 55*, 1943–1948.

282. Ingram, L. O., Conway, T., Clark, D. P., Sewell, G. W., and Preston, J. F. (1987) Genetic engineering of ethanol production in Escherichia coli, *Appl. Environ. Microbiol. 53*, 2420–2425.

283. Ingram, L. O., Gomez, P. F., Lai, X., Moniruzzaman, M., Wood, B. E., Yomano, L. P., and York, S. W. (1998) Metabolic engineering of bacteria for ethanol production, *Biotechnol. Bioeng. 58*, 204–214.

284. Beall, D. S., Ohta, K., and Ingram, L. O. (1991) Parametric studies of ethanol production from xylose and other sugars by recombinant Escherichia coli, *Biotechnol. Bioeng. 38*, 296–303.

285. Jarboe, L. R., Grabar, T. B., Yomano, L. P., Shanmugan, K. T., and Ingram, L. O. (2007) Development of ethanologenic bacteria, *Adv. Biochem. Eng. Biotechnol. 108*, 237–261.

286. Yomano, L. P., York, S. W., and Ingram, L. O. (1998) Isolation and characterization of ethanol-tolerant mutants of Escherichia coli KO11 for fuel ethanol production, *J. Ind. Microbiol. Biotechnol. 20*, 132–138.

287. Underwood, S. A., Zhou, S., Causey, T. B., Yomano, L. P., Shanmugan, K. T., and Ingram, L. O. (2002) Genetic changes to optimize carbon partitioning between ethanol and biosynthesis in ethanologenic Escherichia coli, *Appl. Environ. Microbiol. 68*, 6263–6272.

288. Gonzalez, R., Tao, H., Purvis, J. E., York, S. W., Shanmugam, K. T., and Ingram, L. O. (2003) Gene array-based identification of changes that contribute to ethanol tolerance in ethanologenic *Escherichia coli*: Comparison of KO11 (parent) to LY01 (resistant mutant), *Biotechnol. Prog. 19*, 612–623.

289. Alper, H., Moxley, J., Nevoigt, E., Fink, G.R., and Stephanopoulos, G. (2006) Engineering yeast transcription machinery for improved ethanol tolerance and production, *Science 314*, 1565–1568.

290. Lin, Y. L., and Blaschek, H. P. (1983) Butanol production by a butanol-tolerant strain of Clostridium acetobutylicum in extruded corn broth, *Appl. Environ. Microbiol. 45*, 966–973.

291. Hanai, T., Atsumi, S., and Liao, J. C. (2007) Engineered synthetic pathway for isopropanol production in Escherichia coli, *Appl. Environ. Microbiol, 73*, 7814–7818.

292. Hopwood, D. A., Malpartida, F., Kieser, H. M., Ikeda, H., Duncan, J., Fujii, I., Rudd, B. A., Floss, H. G., and Omura, S. (1985) Production of "hybrid" antibiotics by genetic engineering, *Nature 314*, 642–644.

293. Mansoor, W., Gilham, D. E., Thistlethwaite, F. C., and Hawkins, R. E. (2005) Engineering T cells for cancer therapy, *Br. J. Cancer 93*, 1085–1091.

294. Koffas, M., and Cardayre, S. D. (2005) Evolutionary metabolic engineering, *Metab. Eng. 7*, 1–3.

295. Raab, R. M., Tyo, K., and Stephanopoulos, G. (2005) Metabolic engineering, *Adv. Biochem. Eng. Biotechnol. 100*, 1–17.

296. Bro, C., Regenberg, B., Forster, J., and Nielsen, J. (2006) In silico aided metabolic engineering of Saccharomyces cerevisiae for improved bioethanol production, *Metab. Eng. 8*, 102–111.

AUTHOR INDEX

SUBJECT INDEX

Advances in Enzymology and Related Areas of Molecular Biology, Volume 76
Edited by Eric J.Toone   Copyright © 2009 by John Wiley & Sons, Inc.